HOW PI CAN SAVE YOUR L

USING MATH TO SURVIVE PLANE CRASHES

ZOMBIE ATTACKS, ALIEN ENCOUNTERS

AND OTHER IMPROBABLE REAL-WORLD SITUATIONS

CHRIS WARING

Published in the US by:
ULYSSES PRESS
PO Box 3440
Berkeley, CA 94703
www.ulyssespress.com

First published in Great Britain in 2020 as *An Equation for Every Occasion* by Michael O'Mara Books Limited

ISBN: 978-1-64604-193-0
Library of Congress Control Number: 2021931512

Printed in the United States by Kingery Printing Company
10 9 8 7 6 5 4 3 2 1

Acquisitions editor: Claire Sielaff
Managing editor: Claire Chun
Editor: Barbara Schultz
Proofreader: Joyce Wu
Production: Jake Flaherty, Yesenia Garcia-Lopez
Interior design: Design23
Illustrations: Neil Williams

Contents

Introduction

Equations and formulas (or should that be formulae?). Most of us will be familiar with them from math and science lessons at school. But perhaps after you learned a few to get you through your exams, they now lie quietly in the recesses of your mind, largely forgotten and seemingly unimportant in your grown-up life. After all, you can get through most days with some basic arithmetic and—in times of absolute dire need (perhaps the week before payday)—the calculator on your smart phone. If there is an equation you really need, there's probably an app for it, or a spreadsheet, or a piece of software that does it for you. Why would you want to revisit these wretched, useless, unnecessary things now?

As far as we can tell, our universe follows some rules. We call those rules science and we write down these rules in the language of mathematics. In the language of mathematics, these rules are equations. Everything, whether it's the formation of galaxies or the pattern of freckles on a child's nose, follows the outcomes of these equations. Whether you like it or not, whether you're a fly-by-the-seat-of-your-pants kind of person, or a stickler for order and detail, every aspect of your life is governed by equations.

They don't care whether you understand them or not; they're still in charge of what happens around you. So, perhaps it's about time you and the world of mathematics became a bit more familiar?

Of course, while equations can help you work out a safe distance at which to drive your car to avoid causing a pileup during rush-hour traffic, they can also be very helpful in more extreme situations where there's more on the line than your insurance premium. Imagine that, instead of shuffling off to your pen-pushing job for Mr. Silktie this morning, you were in charge of intercepting the latest message from beings from another galaxy? Or it were down to you to curb a disastrous oil spill in the Pacific before it causes an international incident? These globally important and diplomatically precarious situations would require the humble equation. Mathematics is what drives the world, and improving our mathematics is what will drive our technology farther—and could save the world from a life-threatening energy crisis!

But before we get on to saving lives with mathematics, let's first remind ourselves of the basics. You're going to need these if you want to follow the chapters in this book without a sense of total inadequacy.

We all need help with mathematics sometimes. Even geniuses like Isaac Newton and Albert Einstein struggled at times to get their ideas written down mathematically, and both had help and guidance from mathematicians. While I can't be there to help you as you read, I have written some notes on things you may have forgotten since school that will really help in this book.

Depending on your level of mathematical confidence, you could skip this section initially and refer back to it later when you've realized you've overestimated your abilities.

Order of operations

Whenever you are presented with a string of calculations—or operations, as mathematicians call them—there is an order of priority. Mathematics, unlike the written word, does not always proceed from left to right. Instead, we perform each type of operation in a specific order. This order of priority is often given the acronym BIDMAS:

Brackets

Indices

Division

Multiplication

Addition

Subtraction

For example, $5 - 3 + (2 \times 8) \div 4^2$ contains each of the six elements of BIDMAS. Starting with the brackets, we see that $2 \times 8 = 16$, so our calculation becomes

$$5 - 3 + 16 \div 4^2$$

Next on the agenda is indices, which are also known as powers, and we can see just such a critter nestled above the 4 there. 4^2 means 4 times itself and $4 \times 4 = 16$, so

$$5 - 3 + 16 \div 16$$

Next is division, and 16 ÷ 16 = 1. So then our calculation becomes

$$5 - 3 + 1$$

Adding -3 and 1 gives -2:

$$5 - 2$$

And so, finally, we are left with the rather more straightforward sum of

$$5 - 2 = 3$$

Simplifying fractions

The Equivalence of Fractions is an important concept that states that fractions may have the same value even if they have different numbers in them. For example, we know that one-half is the same as two-quarters:

$$\frac{1}{2} = \frac{2}{4}$$

It is usual to leave fractions in their simplest form, which means using the smallest possible denominator (bottom number) that still gives us a whole number for the numerator (top number). If I didn't know that two-quarters are the same as a half, I could simplify it by searching for a number that divides into both the numerator and denominator. For two-quarters, this number would be 2, as it goes into 2 and 4. If I divide both numbers by 2, the fraction maintains its value, but is now simplified.

If I had eight-twelfths, I could divide these by 2 or 4. I'd use the larger number, as this will fully simplify the fraction:

$$\frac{8}{12} = \frac{8 \div 4}{12 \div 4} = \frac{2}{3}$$

There is no number that divides into 2 and 3, so our work here is done.

Powers and roots

We saw an example of a power in the Order of Operations section. Powers are a shorthand for showing that a number has been multiplied by itself a number of times. For instance, I could write $3 \times 3 \times 3 \times 3 \times 3$ as 3^5. The actual value of 3^5 is 243, which is very different from $3 \times 5 = 15$, the common mistake that is made with powers.

Roots are the opposite of powers. We are most familiar with square roots, which do the opposite—or inverse, as mathematicians like to say—of squaring a number (multiplying by itself once). For instance,

$$8^2 = 8 \times 8 = 64$$
$$\sqrt{64} = 8$$

I squared 8 and square rooting the result of this brought me back to where I started. Just as I can use powers other than 2, I can find roots other than the square root. For instance, if I cube 8,

$$8^3 = 8 \times 8 \times 8 = 512$$
$$\sqrt[3]{512} = 8$$

Solving equations

Equations are, essentially, missing number problems. For instance, if I tell you that my number, multiplied by 4 and with 3 added gives 13, can you tell me my number? One way to write this problem more concisely is to use algebra. If we represent my mystery number with the letter y, then I can write the sentence as:

$$4 \times y + 3 = 13$$

As a shorthand and to avoid confusing the multiply sign with the letter x, $4 \times y$ is abbreviated to 4y:

$$4y + 3 = 13$$

To solve the equation, we start with the answer and do the inverse of each operation in reverse order to find the unknown quantity. In this case, I would start with 13, subtract 3 and then divide by 4:

$$y = (13 - 3) \div 4$$

Notice I have put brackets around the first operation to indicate that we need to do the subtraction first, rather than the division that BIDMAS would dictate. So,

$$y = (13 - 3) \div 4$$
$$y = 10 \div 4$$
$$y = 2.5$$

Equation solved! Alternatively, you can do inverse operations in stages. This is helpful when the unknown is in more than one place in the equation.

$$3a + 6 = 7a - 2$$

For example, if I add 2 to both sides of this equation, I effectively get rid of the -2 on the right-hand side. My equation now looks like this:

$$3a + 8 = 7a$$

I can then subtract 3a from both sides:

$$8 = 4a$$

And finally divide both sides by 4 to get the correct answer:

$$2 = a$$

This method works beautifully for linear equations like the ones above, where the unknown has no power. Quadratic equations—where the unknown is squared—can be harder to solve as they can have two, one or even no solutions. There are various techniques for solving them, but throughout this book I will spare you the detail. So you don't think, though, that these involve magic or some sleight of hand, here is a formula for solving them.
For $ax^2 + bx + c = 0$,

$$x = \frac{-b \pm \sqrt{b^2 - 4ac}}{2a}$$

I leave it as a challenge to the more conscientious reader to verify my results!

Formulae

A formula is a way of showing a mathematical relationship between quantities using algebra. For example, a foot is 30.48 centimeters. I can show this as a formula:

$$c = 30.48f$$

The letter f represents the number of feet and c represents the number of centimeters. If I was working in the US (where the foot is still a standard unit of length), I could use this formula to work out how many centimeters 6 feet is by replacing f with 6:

$$c = 30.48 \times 6$$
$$c = 182.88$$

So, 6 feet is 182.88 cm.

In the example above, c is called the subject of the formula. If you knew how long something was in centimeters but wanted to convert it to inches, you would need to change the subject of the formula to f—that is, you would need to rearrange it to say "f =." This is very similar to solving an equation. I can see that f has been multiplied by 30.48 to give me c, so c divided by 30.48 should give me f:

$$f = c \div 30.48$$

So, if I wanted to know how many feet 182.88 cm was, I would divide it by 30.48 to find it is 6 feet.

Inequalities

Often the aim in mathematics is to be certain, to show that x equals a particular value or values. Sometimes it is either impossible or undesirable to use such certain terms and we want to consider a range of values. This is where we use inequalities. For instance, I know from experience that my family eat more than 7 but up to 12 roast potatoes at every Sunday dinner. If the number of potatoes is represented by p, then $p > 7$ means "p is greater than 7." I think of the inequality symbol as the mouth of a greedy crocodile that always eats the larger of the two things it is offered, in this case p. I can also write this the other way round: $7 < p$, as "7 is smaller than p" has the same meaning as "p is greater than 7." "Up to 12" means that p can be less than or equal to 12, which I can show like this: $p \leq 12$. The extra bar on the symbol means that p can be equal to, as well as less than, 12.

I can show both of these inequalities at the same time, and so indicate the possible range of values for p:

$$7 < p \quad p \leq 12$$
$$7 < p \leq 12$$

This tells me everything there is to know about the roast potato requirements for Sunday dinner.

Pythagoras's theorem

This famous theorem (have you heard of any others?) concerns the sides of right-angled triangles and the relationship between them:

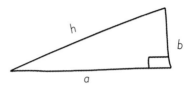

The square of the longest side of the triangle (a.k.a the hypotenuse) is equal to the sum of the squares of the other two sides. If you know the length of the shorter sides and want to know the hypotenuse, this form is useful:

$$h = \sqrt{a^2 + b^2}$$

If you want to know the length of one of the shorter sides, use this:

$$a = \sqrt{h^2 - b^2}$$

Expanding brackets

Sometimes equations feature brackets. If I had a number and said that it, multiplied by a number 4 larger, gave a total of 45, you could write it as an equation like this:

$$n \times (n + 4) = 45$$

This would usually be written without the multiplication sign:

$$n(n + 4) = 45$$

To solve the equation, I need to do something with the brackets. It can help to think of this like a rectangle. If one side of the rectangle is n meters long, and the other is n + 4 meters long, it would look like this:

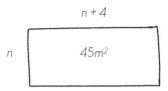

I can split the long side into two sections, one n meters long, the other 4 meters long:

I can now work out the area of each part:

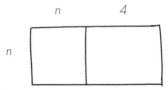

So, the total area of the rectangle is $n^2 + 4n$, which must still make 45:

$$n^2 + 4n = 45$$

Look—no brackets! This process is called "multiplying out" or "expanding" the brackets. This is a quadratic equation, which can be solved using the formula in the Solving Equations section.

Factoring

Factoring is an algebra technique that can be useful when solving equations or rearranging formulae. It is the opposite of expanding brackets. Follow this example with numbers:

$$4 \times 3 + 5 \times 3 = (4 + 5) \times 3$$

If I evaluate both sides of this, I get:

$$12 + 15 = 9 \times 3$$

$$27 = 27$$

So, it's true, and it holds true that I could swap out the three for any other number—four lots of something plus five lots of the same thing will be nine lots of that thing. If, instead of saying "thing" I use a letter, it becomes algebra and the first stage would look like this:

$$4a + 5a = (4 + 5)a$$

Both sides of the equation here are, of course, equal to 9a. This process is called factorizing. I can go farther and replace the 4 and 5 with unknowns too:

$$Xa + Ya = (X + Y)a$$

This is really helpful if you are trying solve an equation where the unknown you are trying to find is in more than one place.

Hopefully, the sections above have activated some memories and you are now ready to tackle the first scenario. Don't worry if you are still feeling a bit wobbly—we will go through the problems in every chapter step by step, with plenty of explanation at each point. And there's no test at the end. You might not have thought about them much since school, but in the pages that follow, you will see that there really is an equation for every occasion.

CHAPTER 1

Born to Louvre

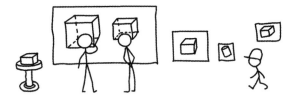

Your reputation as a private security consultant is second to none, and some of your more high-profile cases have recently caught the attention of the international media. When a woman dressed in the height of Paris chic arrives at your office, you quickly clear your schedule. You offer her a drink and agree to help her determine which member of her museum staff is stealing masterpieces and replacing them with nearly identical forgeries. The Louvre has a tight budget in terms of security, and together you must work out how to monitor the newly opened Salle d'Art Mathématique exhibition with as few guards as possible. She requires every part of the room to be within the line of sight of at least one guard at all times to keep the various paintings, sculptures and other artworks safe. How can you do this in the most efficient way possible?

This will require a bit of mathematical logic and geometrical reasoning. If you start to think about the room and the guards in mathematical language, you can represent the number of guards needed as g and then see whether you can narrow down the value of g at all. To do this, you're going to need to consider polygons.

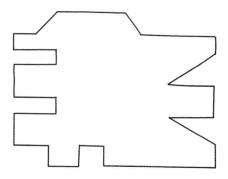

Polygons are two-dimensional shapes with straight sides. The floor plan of most rooms comprises polygons and traditionally involves mostly right angles, though this isn't always the case, as we can see with the floor plan of the gallery.

Polygons are described according to the number of sides they have. A triangle is a polygon with three sides, which happens to be the lowest number of sides possible for a polygon. If you stick two triangles together, edge to edge, you can make a four-sided shape known as a quadrilateral. Quadrilaterals include rectangles, squares, trapezoids, kites, parallelograms and rhombuses. Add another triangle and you have a pentagon, a five-sided polygon. If you keep adding triangles, you add additional sides to your polygon.

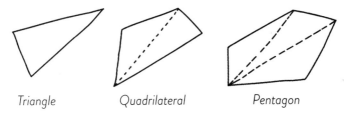

Triangle   *Quadrilateral*   *Pentagon*

Polygons can be convex or concave. Convex polygons are ones where the interior angles at the corners of the shape are all less than 180°. This means that, if you are looking at the polygon from outside, its edges would appear to extend outwards towards you, which is what convex means. Essentially, the pointy bits would stick out. With concave polygons, at least one of the interior

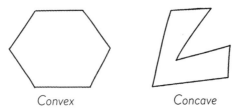

Convex   *Concave*

16

angles is more than 180°, which means there would be bits that, from the outside, appear to point inwards.

Imagine yourself standing inside a room with a convex polygon floor plan. You would be able to see everywhere in that room, no matter where you stood. Mathematically, if you stand at any point in the room, you could draw a straight line from yourself to anywhere else in the room. In this context, the line represents your eyesight. Therefore, one guard can supervise any convex room.

Sadly, perhaps in an attempt to be artistic or maybe just to increase hanging space, the Salle d'Art Mathématique is a concave polygon with twenty-eight sides—an icosioctagon, to be precise. There is no point inside it where you can draw a straight line to every part of the polygon, so we can definitely say that more than one guard is required to watch the entire room. So, we now know that g > 1. Perhaps a little obvious, but we've got our starting point.

We know that you can build polygons out of triangles. As you might also recall from school, the angles within a triangle always add up to 180°. As triangles have three angles, each angle must be less than 180°, which means that triangles cannot be concave. Therefore, no matter the triangle, a single guard can always see the whole of a triangular room. This is not true for quadrilaterals or polygons with more than three sides, of course, as they could be concave. So, you now know that your client will require, at most, one guard for every triangle that makes up the polygon. It might be useful at this point to mention that a polygon is always made up of two fewer triangles than it has sides: a triangle has three sides and (obviously) one triangle, a quadrilateral has four sides and contains two triangles, a pentagon has five sides and three triangles, and so on.

So, for a room with n sides, g would need to be at least $n - 2$, which gives me $g \le n - 2$. If you combine this with our previous statement for g, you will see that

$$1 < g \le n - 2$$

For the gallery predicament, where n = 28, you can see that this becomes

$$1 < g \le 26$$

So, one way of splitting the room into triangles would look like this:

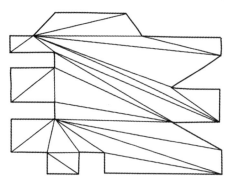

There are other ways, of course, but you can be sure that a 28-sided polygon will have 26 interior triangles.

The client seems happy with your logic, but has some understandable concerns about having twenty-six guards loitering in her gallery, even if her budget could stretch to it. But you can reassure her that your work is not yet finished, and with a bit more effort you can bring this number down considerably.

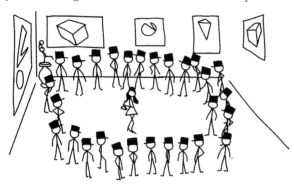

Let's think about what would happen if you placed the guards in a corner of each of the triangles. If we label the corners of each triangle A, B or C, you will find it is possible to do it in a way that no two corners with the same letter are adjacent to each other.

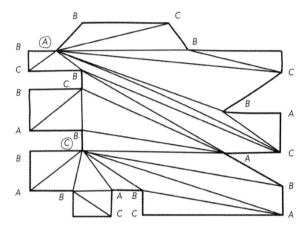

Why is this important, you may well ask? Well, it means that if you pick a letter—A, B or C—and only post a guard at each corner that has that particular letter, then every triangle still has a guard, but some guards will be covering more than one triangle. This is because many of the corners are shared by more than one triangle. For this reason, when you cut the room into triangles, you tried to have as many triangles share corners as possible, specifically at the two circled corners. So, if the polygon has n corners, there must be about n ÷ 3 of each of A, B and C corners. Since n ÷ 3 will not necessarily be a whole number, we always round down to the nearest whole number, known as the floor function in mathematics. Depending on the exact shape of the room, there may be ways to reduce this number farther, but you can be confident that this value will be the greatest number of guards needed. So, you now know that

$$1 < g \leq \lfloor n \div 3 \rfloor$$

The brackets with their tops cut off represent the floor function. For our job, we can see that

$$1 < g \leq \lfloor 28 \div 3 \rfloor$$
$$1 < g \leq \lfloor 8\tfrac{2}{3} \rfloor$$
$$1 < g \leq 8$$

This tells you that the most guards required for a 28-sided room is eight, no matter the shape of the room. At less than a third of the previous figure, the client seems much happier with this. But can you do better?

You now start looking for heuristic solutions. These are solutions that rely on spotting things to improve your solution for the particular problem you have. There is no guarantee that your solution is perfect with this method, but anything that reduces the number of guards required will please your client.

Elegant math

The Art Gallery theorem was first proved by Czech-Canadian mathematician Vaclac Chvátal in the 1970s and then later simplified by American mathematician Steve Fisk, who came up with the geometric proof used in this chapter. The idea came to him when he was snoozing on a bus in Afghanistan and is considered by mathematicians to be "elegant," which means that a complex problem is broken down into a proof that even non-mathematicians can follow. For this reason, the proof is included in The Book, a compendium of the most beautiful proofs in all fields of mathematics, written by Martin Aigner and Günter Ziegler.

You notice that there are a few places on the floor plan where a guard could see most of the room. For instance, a guard standing at the corner shown below would be able to see all of the room apart from the shaded areas:

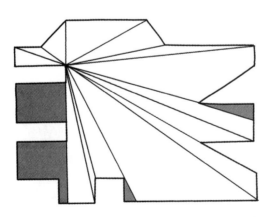

You can then place another guard to try to cover these areas; for instance, a second guard placed as below could see most of the areas required:

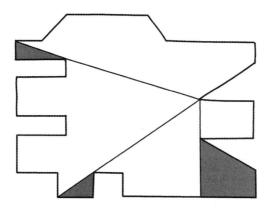

This leaves a tiny blind spot in the bottom left-hand corner that a third guard could cover.

Your client seems pleasantly surprised by the simplicity of your reasoning and the low number of guards she must employ. She reaches into her Louis Vuitton to pay your fee. "Unnecessary, Madame," you say. "I shall happily waive my fee in return for a lifetime membership to the Louvre and a private viewing of the Salle d'Art Mathématique."

"Of course, Monsieur," she replies, her mind already calculating the costs of hiring the three guards.

Obviously, though, you'll wait until the thief has been caught before you make the trip.

Dodgy business

Art theft and forgery is, and always has been, big business. Even Michelangelo got started in the art business with a bit of forgery. He was something of an amateur compared to some of the master art criminals we know about. The Greenhalgh family from the UK made £1 million selling fakes produced in their garden shed between 1989 and 2006. Shaun Greenhalgh has been labeled the most diverse forger ever, working in several different paint media, as well as sculpture and metalwork. He was self-taught and produced at least one hundred and twenty forgeries. The family were finally caught in 2006, not because an expert saw through a forgery, but because they used the same letters of authenticity for several works. Stéphane Breitwieser from France stole over twon hundred fifty artworks from smaller museums and galleries while traveling around Europe between 1995 and 2001. With his girlfriend keeping watch, he would remove paintings from their frames and escape. He made no attempt to sell any of the works, and sadly many of them were destroyed by his mother after he was finally captured in the act of stealing a 500-year-old bugle from a museum in Switzerland. In 1990 a team of criminals stole art valued at $500 million from a museum in Boston, including paintings by Vermeer, Degas, Rembrandt and Manet. None of the paintings have been seen since, no one was ever arrested and there is still a $10 million reward offered for their safe return.

CHAPTER 2

Phone Home

The Search for Extraterrestrial Intelligence (SETI) Institute, which looks for evidence of alien life and civilizations, was founded in 1984. Part of its important work involves searching for incoming radio signals from space. You have won a competition advertised in your local newspaper and are interning there before starting your astronomy degree. What luck then, that on your first day a signal from what appears to be an alien civilization is detected! Shadowing one of the senior astronomers, you are given the opportunity to work with her to pinpoint the location of the signal's origin. You know from the parameters of the telescopes used when the signal arrived that the location must be within an area bounded by four stars. If you can calculate the area between the stars, you can then set your telescopes to search in a methodical manner.

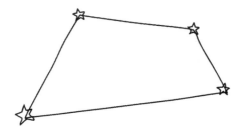

Space is big! So big that astronomers rarely use standard units of length like meters or kilometers. The preferred unit is called a parsec. As the Earth moves around the sun, the distance between stars appears to change, depending on how far away they are. Think of closing one eye at a time. The view of more distant objects doesn't appear to change when you do this, but your hand held near your face appears to move as you swap eyes. The distance between your eyes is not great and won't affect your view of the stars, but the distance across Earth's orbit around the sun is about 300 million kilometers—enough to have a measurable impact on the view of the stars. This phenomenon is called parallax, and the parsec is defined as the distance to an object that has a parallax of one three-thousand-six-hundredth of a degree when viewed from Earth from each side of the sun. That's quite tricky to wrap your head around, so the astronomer advises you to use light-years instead. A light-year is not a length of time, but rather the distance that light can travel in a year. As space is essentially empty apart from a few bits of dust, the occasional solar system and a few hydrogen molecules, it's important that you use the speed of light in a vacuum, since it moves more slowly through other materials like glass or water.

In the same way that space is big, light is fast. Albert Einstein showed that the speed of light is the speed limit of the universe—nothing can travel faster. We know that the speed of light is 299,792,458 meters per second (m/sec) in a vacuum. That's just under 300,000 kilometers per second (km/sec). Traveling at this speed, you could travel from Earth to the moon and back in under three seconds.

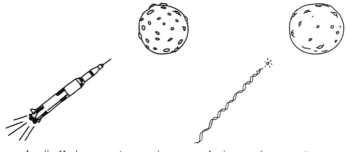

Apollo 11: Journey time to the moon—3 days, 3 hours, 56 minutes

A photon: Journey time to the moon—1.3 seconds

To help you appreciate the vastness of space and the scale of SETI's search, the astronomer asks you to work out how far a light-year is in kilometers. You can work it out like this:

300,000 kilometers per second (× 60)

= 18,000,000 kilometers per minute (× 60)

= 1,080,000,000 kilometers per hour (× 24)

= 25,920,000,000 kilometers per day (× 365)

= 9,460,800,000,000 kilometers per year

That's 9.5 trillion km. To put this into context, the nearest star to the sun is Proxima Centauri and it's about 4.2 light-years away— about 40 trillion km. Our fastest space probe can achieve about 250,000 km/h. Therefore, you can calculate that it would take a mere 160 million hours to get there, or a smidge over 18,000 years.

You now appreciate that, even if we do detect a signal from aliens, traveling to meet them is something that will need some serious improvements in our interstellar travel times.

The astronomer now gives you the star map, conveniently on a light-year grid. It shows the positions of the four stars and the region between them. The shape made by the four stars is a quadrilateral, but not one that has a convenient area formula. You realize that you could try to split it into rectangles and triangles and work out the area from that. We know that the area of a rectangle is its length multiplied by its width: A = lw. Similarly, it's useful to know that any triangle is half the size of the rectangle it fits into:

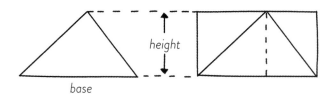

Length and width don't make much sense for a triangle, so we tend to use base and height instead: A = ½bh.

The quadrilateral the stars make is an awkward one to split up into convenient triangles and rectangles, but the area outside the shape is not too bad, and you could work out the area of the quadrilateral by working out the area of the whole map and then subtracting the external shapes. You split the map up as follows:

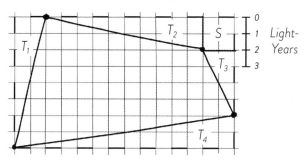

The area of the dotted rectangle is 14 × 8 = 112 and the units will be square light-years, ly^2. You then begin working out the area of the four triangles. The area of T_1 is $\frac{1}{2} \times 8 \times 2 = 8$ ly^2. Using similar calculations, you calculate the areas of T_2, T_3 and T_4 as 10, 4 and 14 ly^2 respectively. The area of the square, S, is 4 ly^2. This means the area of the quadrilateral is $112 - 8 - 10 - 4 - 14 - 4 = 72$ ly^2.

As a conscientious scientist, you wish to check your work using a different method. Fortunately, you remember that there is a really nifty way to do this. In 1899 Austrian mathematician Georg Pick published what is now known as Pick's theorem. This is a way to calculate the area of shapes where the corners lie on points of a grid. The theorem says

$$\text{Area} = i + \frac{b}{2} - 1$$

The letter i stands for the number of interior points and b stands for points on the boundary of the shape.

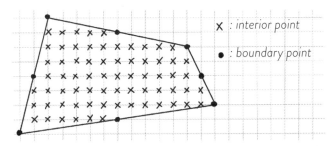

You can see from this diagram that there are 69 interior points and 8 on the boundary, which gives us

$$\text{Area} = 69 + \frac{8}{2} - 1$$

$$\text{Area} = 69 + 4 - 1$$

$$\text{Area} = 72 \text{ ly}^2$$

Proud of your verified result, you give the answer to your mentor, who begins the focused scan for the exact source of the message.

Many scientists think that any message from aliens would be mathematical in nature. The Pioneer 10 and 11 spacecraft, launched in the 1970s, both carried a plaque showing a picture of male and female humans along with information about the location of our solar system, all using the hydrogen atom as a reference for a unit of length, in case the craft met any aliens on their travels. In 1974 a message written by Frank Drake and Carl Sagan was sent from the Arecibo radio telescope in Puerto Rico, containing information about DNA, humanity and the planets of the solar system, mathematically encoded in the signal. The message will take 25,000 years to reach its intended target.

The assumption is that any alien life form we could communicate with would have to have a sufficient understanding of mathematics to have developed the technology to intercept our messages. Using mathematics as a basis for setting up a method of communication would work, but there's always the chance that aliens will have learned all about us from watching the numerous television shows we have broadcast into space over the years. Whatever the reply might say, we can be fairly certain that mathematics will be involved in it somewhere.

Drake equation

Frank Drake is an American astronomer who helped to start the search for extraterrestrial intelligence long before SETI was founded. He also developed a famous equation to attempt to work out how many alien civilizations there should be to contact in our galaxy. It looks like this:

$$N = R_* f_p n_e f_l f_i f_c L$$

R_* denotes how many stars are formed in the galaxy on average each year; f_p is the fraction of those stars that have planets; n_e is the number of planets these stars have; f_l is the fraction of the planets that are predicted to develop life; f_i is the fraction of the life-bearing planets that may develop intelligent life; and f_c is the fraction of the intelligent life planets where the life will emit detectable signals into space. L is the number of years the intelligent life will be broadcasting signals for us to detect. Multiply these together and you get N—the number of alien civilizations we could contact. The unknowns in the equation get more and more subjective as we go along. The latest estimates give numbers in the range from 0 (we are alone in the galaxy) up into the millions (life is fairly common—let's meet up!).

CHAPTER 3

Zombie Apocalypse!

People all over the world are becoming victims of a violent and mindless zombie army—the numbers are rising every day! After the initial outbreak, there is now widespread panic as people do their best to hide and survive. The internet is down and the power is off. The telephone networks, both landline and mobile, are dead. No help is coming. You are the mayor of the small town of Moddleton, with a thousand inhabitants. In one sense, you're lucky—the town is geographically isolated on a large island in the middle of a very wide river. You've already blocked the two bridges that are the only way onto, or off of, the island. But reports start coming in of a sighting—at least one zombie has somehow made it onto the island. Can you use your knowledge of mathematical modeling to come up with a strategy to

save your town, your citizens and yourself?

Populations—and the way they behave—have been of interest to scientists, mathematicians, economists and politicians ever since humans started to live together in large numbers after the dawn of agriculture more than five thousand years ago. Mathematicians translate the behavior of the population into a system of equations, called a model, that they can feed data into to try to predict the behaviour of the population under certain conditions.

These models can be used for a variety of purposes, for example, to explain the cycle of population growth and crashes of lemmings, as well as to predict crop yields and even voting behavior during elections. Of particular interest, though, are the transmission and effects of disease. As a veteran of many political campaigns yourself, you recognize that modeling the zombie plague as a disease could help you work out the outcomes for your citizens. You also know that these models rely on equations called differential equations. In this context, these are equations that, rather than working out a certain quantity such as the number of zombies in the town, work out the change in that quantity—how the number of zombies in the town changes each day.

Differential equations can be difficult to solve analytically, which means getting an exact answer by using the normal way of solving equations. They typically would involve using calculus, which is advanced mathematics. However, they can be solved numerically without too much difficulty, so you can get results by substituting in numbers to fit your model.

The model you set up relies on your making an estimate of certain numerical details called parameters. In this case, the parameter you need to know is how quickly the zombie outbreak is spreading. Before the lights went out, media reports indicated that zombies could attack and "zombify" an average of two people per day, so you use this as the parameter you need. The actual number of people who get zombified, though, will also depend on the number of people available to infect—the fewer people there are, the fewer people the zombies can encounter and infect. This knowledge lets you write down your first equation, initially in words to help your understanding:

Change in zombie population each day = 2 × number of zombies × fraction of human population remaining

You can change this into

$$\dot{Z} = 2 \times Z \times \frac{H}{1,000}$$

Z and H represent the number of zombies and humans respectively. As there are 1,000 humans to start with, this means H / 1,000 is the fraction of humans remaining. $\dot{Z}$ represents the change in the zombie population. You can farther simplify this equation by removing the multiplication symbols:

$$\dot{Z} = \frac{2ZH}{1,000}$$

You simplify the fraction by dividing the numerator and denominator by 2:

$$\dot{Z} = \frac{ZH}{500}$$

If this is the change in the zombie population, it makes sense that the change in the human population must be the negative version of this—if the zombie population goes up by a certain amount, the human population must go down by the same amount. This gives

$$\dot{H} = -\frac{ZH}{500}$$

Similarly, $\dot{H}$ represents the change in the human population.

So, you have your two differential equations. Now, it's time to run the model. The first day is pretty easy to calculate. Assuming just one zombie is on the island, that zombie will infect two humans. This will make the change in the zombie population ($\dot{Z}$) equal 2 and the change in the human population ($\dot{H}$)–2. It's a good opportunity to check that your equations work:

$$\dot{Z} = \frac{1 \times 1,000}{500}$$

Multiplying the numbers in the numerator gives

$$\dot{Z} = \frac{1,000}{500}$$

Which simplifies to

$$\dot{Z} = 2$$

The next day, it will be slightly harder for the zombies to infect people, as the number of humans available has dropped slightly. So, for day two, with H = 998 and Z = 3, you get

$$\dot{Z} = \frac{3 \times 998}{500}$$

Punching this into your solar-powered calculator gives

$$\dot{Z} = 5.988$$

This is not quite the 6 zombies we would expect the 3 existing zombies to produce, due to the slight reduction in the number of humans. You might argue that it makes no sense to have anything other than a whole number of zombies, but remember: this is a model. What it implies is that, by the end of the second day, the zombies will be most of the way to making six new zombies, but not quite there. It's going to happen though. By the end of day two, you have 8.988 zombies and 991.012 humans remaining. This seems quite a low rate of growth, but as you compute the next few days' numbers, you see things rapidly escalate:

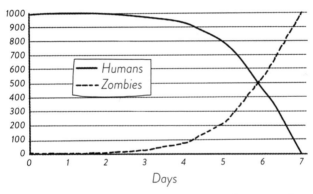

If everyone dies within a week, this would be disastrous for your next election campaign! Obviously, this relies on the human populace standing idly by and letting the zombies do their thing. It does show you that, in a matter of days, things will get very serious, and so you give the order for everyone to stay hidden and barricade themselves in their homes.

How will this affect your model? Well, it will reduce the number of new zombies made each day. If the people are hidden in their houses, the zombies won't have access to them to turn them into zombies. Mathematically, the effect will be to reduce the original 2 multiplier in the equation to a much lower number. Every politician knows that not everyone will do as they are told, whether it's out of stubbornness or necessity to get food or medicine, so you won't be able to make $\dot{Z}$ equal zero. You therefore amend your model to reduce the rate to 0.25—implying that each zombie will make a new one every four days. You run the model and plot the results:

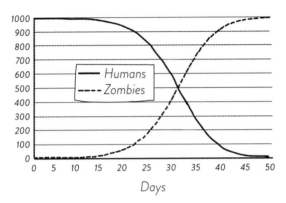

Good news and bad news. The good news is that it takes much longer for the zombies to establish a toehold and really start to do some damage. The bad news is that everyone still gets zombified after fifty days. You can see that this approach will buy you a bit of time. It's clear to you that the only way you will get anyone out of this alive is to beat the zombies at their own game. You decide that you will lock the town down for twenty days, during which time you will conduct some research with your small police force and a few brave military-minded citizens and work out how to conquer the conquerors.

You discover that what the movies show is true—blunt-force trauma to destroy the brain is the best way to defeat a zombie. Fortunately, Moddleton is renowned for its baseball, hockey and golf teams, so there are plenty of bats, sticks and drivers for zombie-smashing. It's also best to hunt them at night when they can't see very well. Hunting zombies is dangerous and increases the likelihood of people getting zombified themselves, but you can't afford to be squeamish in this fight for survival. After a couple weeks of research, you are able to estimate some figures and amend your model.

You need to introduce a third equation—the rate of killing zombies. You estimate that, using your new tactics, improvised weaponry and superior numbers, you can kill 90 percent of the zombies alive each day. This gives the word equation

Change in the number of killed zombies = 90 percent × Z

Using $\dot{K}$ to represent the change in the number of killed zombies and remembering that 90 percent is the same as 0.9, you're left with

$$\dot{K} = 0.9Z$$

While this zombie slaying is occurring, the zombies are still going to zombify some people. You estimate that the rate will increase to 0.75. You also need to consider that the number of killed zombies will reduce the overall number of zombies by the number killed that day. These two effects modify your $\dot{Z}$ equation to

$$\dot{Z} = \frac{0.75ZH}{1,000} - \dot{K}$$

Despite the urgency of the situation, you cannot leave your equation with an improper fraction in it, so you simplify the 0.75 and 1,000 to

$$\dot{Z} = \frac{3ZH}{4,000} - \dot{K}$$

The equation for $\dot{H}$ (the change in the number of humans) remains the same, but as you are hunting them at night, before counterattacking, they've had the whole day to zombify people and so we can't take $\dot{K}$ into account. This gives you

$$\dot{H} = -\frac{3ZH}{4,000}$$

You start to run the numbers. On the twentieth day, there are 920 people and 81 zombies, by your best estimate. So, using $Z = 81$ and $H = 920$ in your three equations gives

$$\dot{K} = 0.9Z$$
$$\dot{K} = 0.9 \times 81$$
$$\dot{K} = 72.9$$

Go humans! On the first day of the rise of the humans, you would expect to kill nearly 73 zombies. This is not without cost, however, as you see when you calculate the change in the number of humans. You substitute in $Z = 81$ and $H = 920$

$$\dot{H} = -\frac{3ZH}{4,000}$$
$$\dot{H} = -\frac{3 \times 81 \times 920}{4,000}$$

Tapping this into the solar-powered calculator gives

$$\dot{H} = -55.89$$

Oh dear! You expect nearly fifty-six people to die in the first night of the counterattack. You have to look at the last equation—the change in the number of zombies—to see whether it was worth it:

$$\dot{Z} = \frac{3ZH}{4,000} - \dot{K}$$

Again, you substitute in $Z = 81$ and $H = 920$ and use $\dot{K} = 72.9$, which you worked out before:

$$\dot{Z} = \frac{3 \times 81 \times 920}{4,000} - 72.9$$
$$\dot{Z} = -17.01$$

You can call me SIR

The real-life model used in this scenario is known as the SIR model, where S stands for susceptible, I for infected and R for recovered. It is the model used in the global COVID-19 pandemic and is based off of early work published by medical doctor Ronald Ross and mathematician Hilda Hudson in the early 1900s, and then subsequent findings by epidemiologist Anderson McKendrick and biochemist William Kermack, who published their own work about a decade later. Just as with zombies, if you can find a way to limit the rate of transmission of the disease, you can "flatten the curve," increasing the length of the pandemic and reducing the strain on health and other essential services. Perhaps if viruses were visible and looked as terrifying as flesh-eating zombies roaming the streets, people would be more cooperative about social distancing.

Yes! You have managed to make the overall change in the number of zombies negative. Despite taking heavy losses on the first night of the counterattack, you have managed to reduce the total number of zombies. From this point on, the zombies' time in Moddleton is limited.

You crunch the rest of the numbers and the graph looks like this:

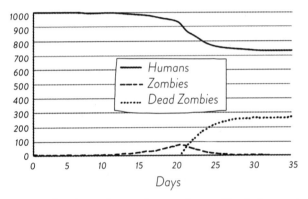

You can see the sharp dip in the numbers of both humans and zombies on day twenty when the counterattack begins, but thereafter things definitely favor the humans. After thirty-five days—five weeks after that first zombie infiltrated Moddleton—you stand on the steps of the town hall with the other 731 survivors and celebrate your victory. What the rest of the world looks like—who knows? After your excellent leadership through this unprecedented crisis, it looks like your position as mayor is very safe indeed.

CHAPTER 4

Easy as Pi

A call comes in on the red phone to your office at Eco Rescue Squad. A ship has run aground near a wildlife sanctuary of endangered sea birds, mammals and aquatic life. It is leaking oil, and the damage to the ship is such that the crew cannot shut it off. Your team has a supply of inflatable tubes called booms with which you can surround the oil to prevents its spreading farther. The crew estimates that the oil is leaking at a rate of 1,000 liters per hour, which has formed a slick 0.002 mm thick with the shape shown opposite from photographs taken by the rescue helicopter. Your team will contain the slick with inflatable booms and then pump the oil and water out, but it will take two hours to reach the site. The boom-deploying ship can travel at 1.7 km/h while laying the booms. If you can accurately estimate the length of boom you'll need, you can send the appropriate crew and tackle the problem as quickly as possible. The prevailing wind and currents in the area are set to remain as they are for some time, so the slick should keep this shape, spreading out from the tanker at an angle of 60°. According to an estimate by the Coast Guard, you have less than twelve hours until the slick reaches the wildlife sanctuary's waters. Can you reach the site and deploy the booms to contain the slick before it reaches the sanctuary?

The oil spill here is roughly the shape of a sector, which is what we call part of a circle. To contain the spill with booms, you have to assume that the shape the booms will make will be a sector, to simplify your calculations. What makes your calculations tricky is that the spill is growing. You need to be able to relate the perimeter of the oil spill to the amount of time that has elapsed.

People have been fascinated with circles for a long time. All circles look the same, regardless of their size—mathematicians say they are *similar*—and therefore always have the same proportions.

Something that has been of particular interest to mathematicians over the ages is that if you take the circumference of a circle (its perimeter, or the distance around the edge of it) and divide it by the diameter (the distance across the circle through the center), you always get the same answer.

The answer is just over three: 3.1415926536 to ten decimal places. The actual number is irrational, which means it cannot be written accurately as a fraction and, as a decimal, it goes on forever without any pattern. We use the Greek letter pi (π) to represent it in its entirety, but have to use a rounded value if we want a numeric answer. Pi crops up in many areas of mathematics and its value seems to define how geometry works in our universe. It certainly helps us work out lengths and areas involving curves and circles. You can rearrange the formula that defines π:

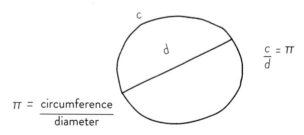

$$\pi = \frac{\text{circumference}}{\text{diameter}}$$

$$\frac{c}{d} = \pi$$

If you multiply both sides by the diameter, you get:

$$\pi \times \text{diameter} = \text{circumference}$$

This allows you to work out the hard-to-measure curved circumference by multiplying the easy-to-measure straight diameter by π. This gives you a formula that may ring a bell from school:

$$C = \pi d$$

If you remember that the radius (the distance from the center of the circle to the edge) is half the diameter, then two radii make a diameter, giving us an alternative formula if we say that $d = 2r$:

$$C = \pi \times 2r$$

Mathematicians normally write this as

$$C = 2\pi r$$

The oil spill is a sector, which is like a slice of a circle made from two radii and a part of the circumference, known as an arc.

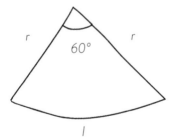

To work out how long the arc is, you need to know how big the sector is, which you can tell from the angle the sector makes. From a satellite image, you are able to see that the angle is 60° and, given that the entire circle would have an angle of 360°, our sector represents a sixth of the total. This means that the length of the arc is one-sixth of the whole circumference:

$$\text{Arc length, } l = \frac{1}{6} \times 2\pi r$$

You make this into one fraction like this:

$$l = \frac{2\pi r}{6}$$

Two over six simplifies to a third:

$$l = \frac{\pi r}{3}$$

The length of booms required is this, as well as the two radii:

$$\text{Boom length required, } b = l + 2r$$

$$b = \frac{\pi r}{3} + 2r$$

You can factor this to get the r in one place:

$$b = r \left(\frac{\pi}{3} + 2 \right)$$

So far, so good. You have a formula for the length of boom you need given the radius of the slick. But, don't feel too pleased with yourself just yet—you still need to factor in that the radius of the slick will increase as more oil flows. Given that it will take some time to get to the site, you need to work out what the radius of the slick will be by the time you arrive.

You used the fact that the slick is shaped like a 2-D sector to help calculate the boom length, but actually the slick is 3-D—like a slice of an incredibly thin cake, 0.002 mm tall. You can work out the volume of the slick by multiplying its area by its depth, and you'll need π again for this. The formula for the area of a circle is given by $A = \pi r^2$. You know we have a sixth of the entire circle, so the volume of the slick will be given by

$$V = \text{Area of sector} \times \text{depth}$$
$$V = \frac{1}{6} \times \pi r^2 \times \text{depth}$$

The depth of the slick is 0.002 mm—about the width of a human hair—which explains why oil spills spread out over such large areas and devastate the environment so badly. We know that 0.002 mm is two-thousandths of a millimeter, and a millimeter is a thousandth of a meter, making the depth two-millionths of a meter. Putting this into your volume formula,

$$V = \frac{1}{6} \times \pi r^2 \times \frac{2}{1,000,000}$$

You put this together as a single fraction:

$$V = \frac{2 \times \pi r^2}{6 \times 1,000,000}$$

You can simplify the 2 over 6 by dividing both by 2:

$$V = \frac{1 \times \pi r^2}{3 \times 1,000,000}$$

This leaves you with

$$V = \frac{\pi r^2}{3,000,000}$$

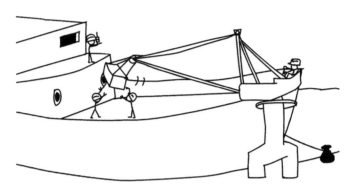

You know that the volume of the slick is increasing by 1,000 liters per hour, which is 1 m³ per hour. This means the volume in meters cubed is this rate multiplied by how many hours have elapsed since the spill began. You set t as the number of hours elapsed, giving $V = 1 \times t$, or simply $V = t$. You put this into your volume formula and start rearranging it to find r:

$$t = \frac{\pi r^2}{3,000,000}$$

Multiplying both sides by 3,000,000 gives

$$3,000,000t = \pi r^2$$

You then divide both sides by π:

$$\frac{3,000,000t}{\pi} = r^2$$

Hairdresser Phil McCory, watching footage of oil-covered otters from the 1989 *Exxon Valdez* oil spill, had a brilliant idea. He took a gallon of oil and poured it into his son's wading pool to simulate an oil spill. He then dropped in a pair of his wife's old tights, stuffed with hair sweepings from his salon. After two minutes the oil had soaked into the hair-filled tights. The International Hair for Oil Spills Program—which aims to collect hair from salons, animal groomers and sheep farmers to use in oil spills—was born. Apparently, human hair is the best because shampooing it allows it to soak up forty times its own weight in oil. The hair can also have the oil squeezed out of it to be reused.

To get r, rather than r², you have to square root both sides of the equation:

$$\sqrt{\frac{3{,}000{,}000t}{\pi}} = r$$

Substituting this into your boom formula, you can tell how much boom you need given how long it's been since the oil spill began:

$$b = \sqrt{\frac{3{,}000{,}000t}{\pi}} \times \left(\frac{\pi}{3} + 2\right)$$

This formula tells you something very useful—that the length of the boom required is proportional to the square root of the time. The boom ship can lay 1,700 m of boom every hour, starting two hours after the spill began. This gives a graph like this:

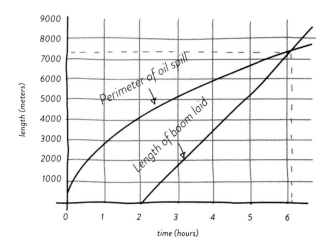

A lot of pi

You may remember being told at school to use various approximations for π when doing calculations—3.14 or 22 ÷ 7 are two common ones. Electronic calculators provide many more digits, which allow us to be accurate enough for almost any use of π you can think of. But, this was not enough for Rajveer Meena of India, who in 2015 broke the world record for memorizing seventy thousand digits of π. It took him ten hours to recite while wearing a blindfold. If memorizing's not your thing, you can always use your computer to calculate more and more digits of π. US cybersecurity analyst Timothy Mullican broke the world record using a home-built computer, finishing in January 2020 after 303 days, calculating 50 trillion digits.

You see that, initially, the oil spill's perimeter grows very quickly, but as time goes by, the rate of growth reduces, giving your team a chance to catch up. By looking at the graph you can see that the amount of boom laid outstrips the perimeter of the oil spill after just six hours, requiring between 7,000 and 7,500 meters of boom.

You breathe a sigh of relief as you realize it should be possible to reach the site, deploy the booms and protect the sanctuary from harm. You load up the ship and set sail.

CHAPTER 5

Bone of Contention

Weeks of excavation under the broiling heat of the Saharan sun have finally paid off. Your team has discovered the fossilized bones of a prehistoric hominid that could redefine our understanding of human evolution. The location is so remote that the only way to the site is on foot or by camel. With your running low on food and water, Mother Nature herself throws another cruel twist your way—a huge sandstorm is forecast, and you must evacuate immediately. Anything left behind will be lost beneath the sands. Naturally, the locals charge for carrying packages, but owing to a local tradition, they work out the price according to the longest dimension of the package. The 40-cm-long femur—thigh bone to you and me—of your find has evidence of early bipedal movement and must be preserved. You, however, have only 35 dirham left and they charge 1 dirham per centimeter. Can you use your knowledge of geometry to find the cheapest way to package the bone without having to break it and make it to the safety of some nearby caves before the whole area is covered in sand again?

Being able to pack things efficiently is very important to our civilization—not only to comply with the impossibly small baggage allowance of most airlines, but also for transporting cargo and products around the world. At the time of writing, there is no mathematical algorithm for working out the best way to pack the back of a van with the maximum use of the space available (if you could work one out, you could name your price and go down in history). But when it comes to putting long straight things in boxes, however, it is Pythagoras all the way. Most of us are familiar with this name (and there's a recap in the introduction if not), but did you know that there is a 3-D version of his famous formula too? This is very useful, as it happens, if you're wondering about fitting a long straight bone in a box where the sides of the box are all shorter than the bone itself. Sounds impossible, but let's turn the problem on its head and work out how long a stick can fit in a box with given dimensions.

Let's say the box is a cuboid, x meters by y meters by z meters. For the moment you can stick with the simpler 2-D Pythagorean equation to work it out. You need to find the length of the diagonal across the box, marked as d:

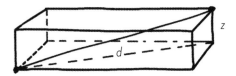

To use Pythagoras you need to know two of the three sides of the triangle, but at the moment you only know one side: z, the height of the box. But, you can find out the length of the side that goes across the bottom of the box using Pythagoras as well:

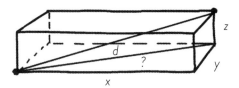

Pythagoras's theorem tells you that this missing side must be this long:

$$? = \sqrt{x^2 + y^2}$$

Square both sides of the equation—you'll see why in a minute:

$$?^2 = x^2 + y^2$$

Back to our original triangle:

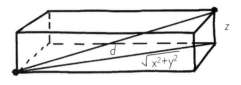

$$d = \sqrt{?^2 + z^2}$$

Now, you can see why you worked out an expression for $?^2$ opposite. If you substitute this in, you get

$$d = \sqrt{x^2 + y^2 + z^2}$$

As you have used unknown values for the sides of your box, you can use this formula for any box, and this is indeed the 3-D version of Pythagoras's theorem.

The Tendaguru fossils

In 1906, German mining engineer Bernhard Sattler found an extraordinary fossil deposit in Tendaguru in Tanzania, then a colony of the German empire. Fossils of enormous dinosaurs were being revealed by erosion on the hillside. He contacted a paleontologist, and eventually nearly 200,000 kg of fossils were excavated from the site over the next six years. The tricky bit was getting the fossils back to Germany for analysis, by first carrying them 60 km overland to the port of Lindi. Pack animals could not be used due to the presence of tsetse flies, which, at the time, carried viruses deadly to cattle, camels and horses. Some of the fossils, once covered in a protective layer of plaster, weighed several hundred kilograms. The solution? Employ an army of locals to carry 4,300 containers on the four-day journey. Even today, some of the containers are still to be opened. The skeleton of a Giraffatitan brancai (a giant sauropod), meaning "titanic giraffe," that resides in Berlin's Museum für Naturkunde came from Tendaguru and is still the largest and tallest mounted skeleton in the world.

Back to the Sahara. It's still a way off, but you spot a rolling cloud of sand blocking out the horizon to one side of your camp and you know the clock is ticking. As you consider the packaging for the priceless artifact, you decide that it makes sense to pack the bone in a cube, where all the sides are the same length. If you have to pay according to the longest dimension, you may as well have the other dimensions be this big as well. Otherwise, you'd reduce the length of d but still pay the same amount of money. So, if all the sides are the same length, then y and z are the same as x. You replace y and z in the formula with x:

$$d = \sqrt{x^2 + x^2 + x^2}$$

$$x^2 + x^2 + x_2 \text{ is } 3x_2$$

$$d = \sqrt{3x^2}$$

The square root of $3x^2$ is the same as the square root of 3 multiplied by the square root of x^2:

$$d = \sqrt{3} \times \sqrt{x^2}$$

You remember that squaring and square rooting are the inverse of each other, and so effectively cancel each other out, giving

$$d = \sqrt{3}x$$

This equation tells you that the diagonal of a cube is $\sqrt{3}$ times bigger than the side. As $\sqrt{3}$ is 1.73 to two decimal places, this means something 73 percent longer than the cube can fit inside the cube.

Rearranging the formula, by dividing both sides by $\sqrt{3}$, gives you this:

$$\frac{d}{\sqrt{3}} = x$$

As the femur is 40 cm long,

$$\frac{40}{\sqrt{3}} = 23.1 \text{ cm}$$

Now, the real-life femur isn't as narrow as a theoretical line, so the box will have to be a bit larger than this, but you can now be certain that big things can fit in small boxes.

You build a crate around the femur, which also has some space around it for some of the other smaller bones your dig has uncovered. You hand over the last of your team's cash and watch the crate, lashed to a camel, ride off into the sunset, before you start to follow on foot.

CHAPTER 6

The Last Train from Vladivostok

The last train of the day, which you desperately need to catch, has just pulled into the station when you suddenly realize that the briefcase containing the evidence against a dangerous rogue agent, representing a year's deep undercover work in Vladivostok, is not in your hand. You've left it at the ticket office, no doubt while you were fumbling through your forged paperwork and travel documents. You look over the tracks and through the double doors of the station building—thankfully you can still see your bag, but your MI6-trained peripheral vision catches sight of a patrolling policeman, who will confiscate the unattended luggage if you don't get to it quickly. You estimate that the policeman is 60 m from the briefcase and walking slowly, at around 1.5 m/sec. You are 200 m away. How fast do you need to run to retrieve your briefcase and catch your train before it leaves the station?

You may remember this handy speed, distance and time triangle from math or science lessons at school. These three quantities are related, and if you know any two of them, you can work out the third. For example, if you wanted to work out how far you might travel (the distance), you cover up D and you'll see that you need to multiply your speed by the time taken. So, from this very simple triangle, we get three incredibly helpful formulae:

Speed = distance ÷ time

Distance = speed × time

Time = distance ÷ speed

It's easy to remember the first formula if you think about driving a car and miles per hour (mph). We know that miles is a distance, hour is a time, per means divide—therefore, speed is distance divided by time.

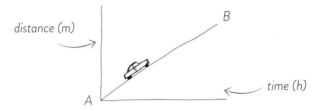

Distance and time are fundamental quantities and have their own units. Scientists and mathematicians favor meters (m) for the unit of distance and seconds (secs) for the unit of time. Speed is a derived (not fundamental) quantity, so its units are meters per second (m/sec). We also commonly use miles per hour in our cars.

Your particular predicament as you stand, frozen by indecision on the platform, has two distinct parts. First, we need to work

---

Fundamental units

The metric system was established by the French Academy of Sciences shortly after the French Revolution in the late eighteenth century and quickly caught on worldwide. It's nice and easy to use—everything is based around powers of 10, as opposed to the previous systems, which require less mental-arithmetic-friendly calculations. *The Système international d'unités* has seven base units from which all the others can be derived. These are the familiar second, meter and kilogram for time, distance and mass, along with the not-so-everyday ampere, kelvin, candela and the amusingly named mole denoting, respectively, electric current, temperature, brightness and number of particles in an atomic mass of substance.

out how long it will take the policeman to reach the briefcase. We know the distance and the speed, and so

$$\text{Time} = \text{distance} \div \text{speed}$$
$$= 60 \div 1.5$$
$$= 40 \text{ secs}$$

The second part of the problem requires calculating how fast you need to run to reach it before him—there's no point running like a madman and drawing attention to yourself in the process if you've time to spare. If we use the 40 secs, this will tell us the speed that will get you to the briefcase at the same time as the policeman:

$$\text{Speed} = \text{distance} \div \text{time}$$
$$= 200 \div 40$$
$$= 5 \text{ m/sec}$$

A speed of 5 m/sec is a slow jog—not too suspicious in a busy train station.

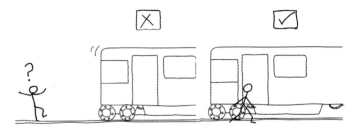

To see whether you can reach the back of the train, you have to work out whether you and the end of the train are the same distance away from the beginning of the platform at the same time. The train has a caboose car that you could climb aboard—if you can catch up with the train. The question is, should you sprint, run or jog after it? The train is 75 m ahead of you, and you know that it will trundle along at 10 km/h until it clears the platform, which is 200 m long.

It might seem an obvious point, but it is important here to make sure we use the same units throughout. If we use meters per second, we need to convert the train's speed of 10 km/h into m/sec: 10 km/h is 10,000 m/h, as a kilometer is 1,000 meters. If the train goes 10,000 m in an hour, you could divide by 60 to find out how far it goes in a minute, and then divide by 60 again to find out how far it goes in a second. If we put all this into a calculation, it'd look something like this:

$$10,000 \text{ m/h} = (10,000 \div 60) \text{ m/min}$$
$$= 166.7 \text{ m/min}$$
$$166.7 \text{ m/min} = (166.7 \div 60) \text{ m/sec}$$
$$= 2.8 \text{ m/sec}$$

We also need to think about the fact that the platform is only so long, and all those afternoons in your apartment eating *medovik* while poring over encrypted documents have done nothing for your personal fitness. The train has 200 − 75 = 125 m of platform left before it will accelerate beyond the speed of human running. How long will that take? Well, we know that time = distance ÷ speed, so

$$\text{Time} = 125 \div 2.8$$
$$= 44.6 \text{ secs}$$

So, whichever mode of running you choose, you must reach the train before 45 secs have elapsed. Let's look at sprinting, running and jogging in turn.

We know the speed of the train (2.8 m/sec) and that initially it has a 75 m head start, but what is your sprinting speed? The 100-meter world record holder, Usain Bolt of Jamaica, can hit a top speed of 12.4 m/sec. Assuming you're not an Olympic athlete as well as an international spy, let's say yours, perhaps fueled by adrenaline and black Russian tea, is 8 m/sec. You can maintain this for about twelve seconds before you would need to stop and take a nap.

Knowing that, we now have all the information we need to start using the distance = speed × time formula. Time is the unknown you want to find, so you will need to use "t" to represent it.

For you,

$$\text{Distance} = 8 \times t$$
$$= 8t$$

For the train,

$$\text{Distance} = 2.8 \times t$$
$$= 2.8t$$

We have to remember that the train has a 75-meter head start, so we'll have to add 75 m to this to see how far it is from the beginning of the platform:

$$\text{Distance} = 2.8t + 75$$

We want to know when the distance you have traveled and that which the train has traveled are the same, so you can equate these two and then solve to find t:

$$8t = 2.8t + 75$$

Subtracting 2.8t gives

$$5.2t = 75$$

Then, you divide by 5.2:

$$t = 75 \div 5.2$$
$$t = 14.4 \text{ s}$$

As you can only sprint for 12 sec, this means that you get close to the back of the train before your speed begins to decrease. You can only watch helplessly now as the train heads off into the sunset and the policeman arrives to ask you some pointed questions …

We've done a lot of the hard work in the previous example. The train's equation will stay the same and the only change will be your running speed. The world record for the 400 meters is held by Wayde van Niekerk of South Africa, and he managed an average speed of around 9.3 m/sec. We'll say your speed is 5.7 m/

sec, which you can keep up for about a minute. So, our equation becomes

$$5.7t = 2.8t + 75$$

Subtracting 2.8t gives

$$2.9t = 75$$

The next step is to divide by 2.9:

$$t = 75 \div 2.9$$
$$t = 25.9 \text{ s}$$

Success! This is within the minute you can run and the 45 sec before the end of the train leaves the platform. You vault balletically onto the back of the train and catch your breath before heading to the restaurant car for a well-deserved shot of celebratory vodka.

Why run and risk attracting attention? Perhaps a more leisurely jog will get you there in time? Assuming you trot along at a weekend-jogger speed of 4.4 m/sec and you can keep this up for much farther than the length of the platform,

$$4.4t = 2.8t + 75$$

Once again, you subtract 2.8t:

$$1.6t = 75$$

Then, you divide by 1.6:

$$t = 75 \div 1.6$$
$$t = 46.9 \text{ secs}$$

Disaster! You have misjudged it, and just as you get to within a few meters of the train, you run out of platform and can only look on as the train speeds off towards Europe. The long jump world record is 8.95 m, but that's a whole different chapter …

When we have to solve several equations as we have in this scenario, it can be simpler to plot a graph showing distance against time. We can see the journeys of the three speeds of

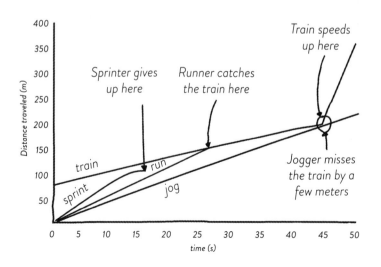

movement and the train on the same graph. If the lines of the runners cross the line of the train, this implies that you and the train are in the same place at the same time.

You can see that the steepness or gradient of each line represents the speed of the object. The sprint line is steepest, followed by the run and then the jog. The train has the shallowest line initially, but it accelerates after 44.6 secs to become the fastest. Note that it also starts at 75 m, when time is 0 due to its head start.

Throughout this chapter we have ignored acceleration—things speeding up and slowing down. This is a fair assumption to make as the train was already moving as you started running for it and it does not take long for people to hit their sprinting, running or jogging speeds. We also get accelerations when things go around bends—that uncomfortable force you feel when driving around a turn. I've assumed that the platform here is nice and straight, so we didn't have to concern ourselves with acceleration. Acceleration does, however, feature heavily in the following chapter.

CHAPTER 7

Lights, Camera, Action!

Your big break in Hollywood has arrived—you land the job as the fight coordinator on the set of *Kung Fu Female*, a new big-budget action movie. The director is famous for his refusal to use any CGI to embellish the action. For this scene, he wants the heroine, Jenny Roundhouse, to deliver her trademark kick to her nemesis and send them flying across the room. You rig the stunt performer with a harness attached to a wire that leads to a heavy sandbag that can be dropped down a lift shaft for a limited distance, pulling the stunt performer horizontally through the air. How heavy should the sandbag be? Too little and the excitement is diminished. Too much and you could find yourself having to work out how to extract a human being from a plasterboard wall.

Before we start, let's talk about weight and mass: despite sounding similar, they are in fact not the same thing. We use them interchangeably in everyday language, which compounds the problem.

Every atom has a certain mass, according to the type and number of subatomic particles that make it up. If you add up the masses of every atom in your body together, you will find your total mass. If you went to the moon, or to a space station in orbit, that mass would not change, as you would still be made up of those atoms.

As you can't work out your mass by sorting and counting all your atoms, we often use weight to help us. Weight is a force caused by gravity and it is directly proportional to your mass, pulling you downwards towards the center of the Earth. The more mass, the more weight.

As we'll see too in Chapter 8, this happens because we are standing on the Earth. Forces can do three things to an object: change their speed, change their direction, or change their shape. When forces are balanced, nothing happens and things stay the same. When you are standing up, your weight is acting downwards but your legs produce a force that cancels it out, so you do not change speed, direction or shape. If your legs stop producing this force, potentially all three could happen. Likewise, when I drive my car at a constant speed, the engine produces a force to make the car accelerate forward, but this is being canceled out by air resistance, so the car remains at a steady speed. Put my foot down, and the engine force increases and the car speeds up. As the car speeds up, it meets more air resistance and eventually the air resistance will increase to match the new engine force. When this happens, the car has reached a new steady speed.

If I took you to the moon, while your mass wouldn't change, your weight would. You would weigh less because gravity on the moon is weaker and you could bounce around like an astronaut. The difference between mass and weight is very important in the equations you'll be using to work out what will happen to your stunt performer.

Isaac Newton (the apple guy) worked this out in the 1600s. His stroke of genius was to realize that human experience is dominated by the force of gravity and to think about what would happen if it were not there. He came up with three laws:

I: If the forces on an object cancel out, the object will remain moving (or not moving) as before

II: Force = mass × acceleration

III: Every force has an equal and opposite force

Newton's second law gives us the equations you need to work out how heavy to make the sandbag.

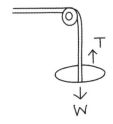

The weight of the sandbag (a force) is given by its mass multiplied by the acceleration due to gravity at the surface of the Earth, which is 9.81 m/sec^2 and is usually represented by the letter g. This gives us the formula for weight:

$$W = mg$$

The other force acting on the sandbag is the tension in the wire, which acts upwards. If the tension in the rope equals the weight of the sandbag, then the sandbag won't move. If the weight of the sandbag is greater than the tension, the overall or resultant force will act downwards and the sandbag will begin to accelerate. For the sandbag,

Resultant force on sandbag = Weight force − Tension force

If the mass of the sandbag is M kg and the tension is T Newtons (as the unit of force is known),

Resultant force on sandbag = Mg − T

According to Newton's second law, force is mass times acceleration, so if I call the acceleration of the sandbag a, I can replace the resultant force with Ma:

$$Ma = Mg - T$$

At the other end of the wire is the stunt performer.

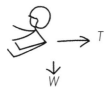

If they time their jump correctly, they will be in the air as the sandbag starts to move and so there will not be any force to oppose the tension in the wire pulling them to the right. The mass of the stunt performer still produces a weight force, but this acts downwards and so won't interfere with the tension force, which will be horizontal. For the stunt performer,

Resultant horizontal force = Tension force

If the mass of the stunt performer is m,

$$ma = T$$

So, you have two equations and you need to know how the acceleration will change as you increase the mass of the sandbag. The value of a must be the same for both equations: the performer gets pulled by the wire as the sandbag falls. You don't know anything about the tension in the wire, so it would be good to eliminate that from your work. As you know $T = ma$, you can replace the T like this and start to find the acceleration:

$$Ma = Mg - ma$$

You add ma to both sides:

$$Ma + ma = Mg$$

Then you factor the left-hand side:

$$(M + m)a = Mg$$

Dividing both sides by the bracket gives

$$a = \frac{Mg}{M + m}$$

If the mass of the stunt performer is 65 kg and g is 9.81 m/sec^2, this gives you

$$a = \frac{9.81M}{M + 65}$$

You can now plot a graph of this to see how the acceleration changes as M increases:

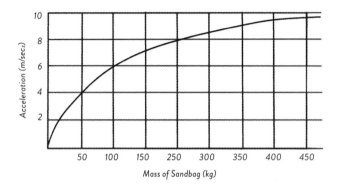

Mass of Sandbag (kg)

You see that, no matter how heavy the sandbag, the acceleration only ever creeps up towards a value just less than 10 m/sec². Thinking about it, this makes perfect sense—there is no way you can get the sandbag to fall faster than gravity, especially when it is dragging the stunt performer behind it. So, the upper limit of the acceleration, even if the sandbag was really massive, would be g: 9.81 m/sec².

Is this an acceptable acceleration for the stunt performer to experience? Well, accelerations are often measured in gs and the human body can tolerate quite high g-forces for limited periods of time. A ride on a rollercoaster could subject you to 3 or 4 gs with no ill effects, so your well-trained stunt performer should be fine.

Bruce Lee's one-inch punch

Legendary martial artist Bruce Lee could punch a grown man from 2.5 cm away and send him flying across the room. Decades of training allowed him to activate most of the major muscles in his body in rapid succession to unleash this explosive power, with his fist hitting the target at 200 km/h, according to his manager. A good way to analyse collisions is to use momentum, which is an object's mass multiplied by its velocity. In collisions, momentum is said to be conserved, which means that the total momentum in the system doesn't change. When Lee punched the man, he transferred the momentum of his fist and arm. We know that Bruce Lee's mass was about 61 kg and that an arm is about 5.3 per cent of someone's total mass on average. So Bruce Lee's Arm Momentum—BLAM for short—would be:

$$BLAM = mv$$
$$BLAM = 61 \times 5.3\% \times 55$$
$$BLAM = 178 \text{ kg m/sec}$$

BLAM gets transferred to the target, which we'll assume is an average Joe with a mass of 75 kg:

$$mv = 178$$
$$v = 178 \div 75$$
$$v = 2.37 \text{ m/sec}$$

This clearly involved a lot of estimation, but we can see that the result is the man moving backwards with some speed.

You now need to work out how far they will travel. We are going to go into more detail concerning how things travel through the air in a later chapter, but let's get a rough idea with what we already know. Let's assume that you use a heavy sandbag that allows you to accelerate the stunt performer at 8 m/sec² and say that they are in the air for one second. The acceleration tells us that, every second, the performer gets 8 m/sec faster. This means that by the end of their second-long flight they will be going 8 m/sec. If you assume that they are not moving at the start, then their average speed for the flight will be 4 m/sec (because the average of 0 and 8 is 4) and hence they will travel 4 m. If the stunt performer can throw themselves backwards at the same time, this distance could be teased out even farther. What if you replaced the sandbag with a winch that could allow acceleration faster than g? How much is too much?

Well, going about your day-to-day business, it would be rare for you to experience more than the 1 g provided by gravity. A decent sports car might get up to 1 g of acceleration. To get much higher than this, you really need specialist equipment, like fighter aircraft, roller coasters, Formula One cars or rockets. People can tolerate high g-forces for short periods of time; on one occasion—a motorsports accident in Texas in 2003—Swedish racer Kenny Bräck just barely survived a crash that peaked at 214 g, thanks to the safety equipment in his car. A trained stunt performer, braced for the acceleration, wearing a supportive harness to distribute the forces on her, could handle considerably more than the 1 g you get from the sandbag. So, you can ask the director how over-the-top they want the kick to be, and if they agree to ramping things up, perhaps invest in a powerful winch.

CHAPTER 8

Worth Your Weight in Gold

As a gold and precious-metal dealer, you fly all over the world to meet your prospectors when they make a notable find. Your Australian contact has struck gold, and you take a flight over to see him, although you'll need to make it quick as you have a meeting with a buyer in Singapore later that day. Then, it occurs to you—could you save some money using your knowledge of physics? You know that the strength of the force of gravity reduces as you get farther away from the Earth. You plan to pick up the Australian and their gold and fly them to Singapore with you, weighing the gold during the flight. As you would be farther from the Earth, the gold would weigh less than it would on the ground, giving you a discount on the amount you would have to pay. So, how much could you save on 100 kg of gold?

Gravity is the force that attracts masses together. It keeps us stuck to the surface of the Earth most of the time, but it also keeps the moon in orbit around the Earth, the planets around the sun, and even makes entire galaxies of stars rotate about their center. In the previous chapter, you used the formula for weight, $W = mg$. This is a bit of a hack, as it only really applies at the surface of the Earth. It is actually derived from Newton's Law of Universal Gravitation (to give it its full title), which looks like this:

$$F = G \frac{m_1 m_2}{r^2}$$

The formula tells us that the force of gravity between two masses is the product of the two masses (m_1 and m_2) divided by the square of the distance between them (r). This is then multiplied by the gravitational constant, G. Any two masses attract each other, but

Rapid weight loss

If you can't afford a private jet to help with your weight-loss plan, simply travel to the top of Mount Huascarán in Peru. Its height (6,768 m) certainly helps in the same way as the plane, but you can take advantage of the fact that the Earth, like so many people seeking weight loss, is a bit bulgy around the middle, due to its spin. As Mount Huascarán is near the equator, the natural bulge of the Earth helps lower gravity to almost the same value as in the plane at 15,000 m. Of course, climbing nearly seven vertical kilometers might help you drop a few pounds too!

because G is such a small number (0.00000000006674 m³/kg s²), we require very large masses to produce a noticeable force. Moons, planets and stars have large enough masses to produce a force that we will notice.

The denominator of the equation is r²—the square of the distance between the two objects. This is where you may be able to save money. You can't reduce the mass of the prospector's gold, nor the mass of the Earth, and the gravitational constant is, as the name suggests, constant. But you can alter the distance between the center of the Earth and the gold if you conduct your business at altitude.

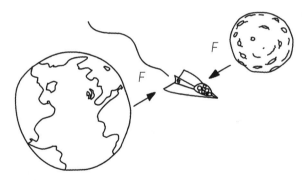

We seem to have two equations for the same thing here—Newton's one and W = mg. How can you tell which one to use? The mass of the Earth is estimated at about 5,972,000, 000,000,000,000,000,000 kg. That's about six yottakilograms or six brontograms, if you ever need to drop a fact bomb at a dinner party and don't want to be invited next time. The distance from the center of the Earth to the surface is 6,371 km on average, or

6,371,000 m. We saw the value of G above, so if I put all those together,

$$F = 0.00000000006674 \times$$
$$\frac{5{,}972{,}000{,}000{,}000{,}000{,}000{,}000{,}000 \times m_2}{6{,}371{,}000^2}$$

$$F = 9.81 m_2$$

G, m_1 and r^2 combine to make 9.81 m/sec^2, which is g. So, on the surface of the Earth, where G, m_1 and r are constants, we can use W = mg. For anything that is not on the surface of the Earth, like your jet, you should use the law of gravitation.

At sea level, the force between the Earth and the 100 kg of gold is given by

$$F_{sea\ level} = 0.00000000006674 \times$$
$$\frac{5{,}972{,}000{,}000{,}000{,}000{,}000{,}000{,}000 \times 100}{6{,}371{,}000^2}$$

$$F = 982N$$

This force of attraction is the weight of the gold. It's worth noting that the gold is attracting the Earth with the same force, but as the Earth is so heavy, the acceleration that this force causes is negligible, and the forces of everything knocking about on the different sides of the Earth tend to cancel each other out anyway. So, when you stand on scales and weigh yourself, you are actually working out the force required to hold you up against gravity. The scales themselves are calibrated to say that every 9.81N is one kilogram.

Now, your bizjet happens to be able to fly at a maximum altitude of 15 km. This increases the value of r by 15,000:

$$F_{15km} = 0.00000000006674 \times$$
$$\frac{5,972,000,000,000,000,000,000,000 \times 100}{(6,371,000 + 15000)^2}$$

$$F = 977N$$

So, the gold's weight is 5 newtons less at this altitude than at sea level. This isn't a huge difference—equivalent to a mass of 500 g, in fact, or a can of beans. It's important to note that how you weigh the gold matters too. If you use an old-fashioned beam balance, with two bowls—one for the gold and one for the masses—your scheme won't work. This is because the masses you use to weigh the gold will also have a lower weight, so the gold will still appear to have a mass of 100 kg—fail! But, seeing as you have a private jet, you can probably also buy a nice high-precision electronic scale.

Let's take a look at the effect on the price. At the time of writing, gold is worth about $54,000/kg. The 500 g is therefore a saving of almost $27,000, which will buy you quite a few tins of beans. You'll still be paying the prospector $5,365,483, so everyone can be happy with this outcome.

CHAPTER 9

Three-Pointer

You look up and see the clock tick down—only seconds remain in the final of the basketball championships. Your team is two points behind, but, if you sink a long shot from where you are, you'll score three points and go down as a high school legend. You know that there are scouts from the professional leagues in the audience, and now is your chance to show them how you can perform under pressure. If you succeed, your dream of a basketball scholarship and a career as a professional player will be one step closer to becoming a reality. Miss, and you will let down the team, your coach and everyone supporting you. You cannot let that happen. How fast and at what angle should you launch the ball?

Objects that are launched and do not have any other means of thrust are called projectiles. Much study has gone into how projectiles behave, as, let's face it, we humans do love to launch stuff around, whether it's for hunting (rocks, spears, arrows, bullets, etc.), recreation (balls) or war (cannonballs, grenades, bullets, etc.). Even the name for the study of projectile motion, ballistics, comes from that of an ancient weapon, ballista—a giant crossbow, which flung spears or rocks at the other lot.

All these things follow a path or trajectory that depends largely on the speed and angle with which they were launched. The trajectory will be in a curved shape known as a parabola, which is defined as the shape you get if you cut through a cone parallel to its sloping side:

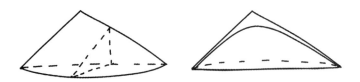

When you look at the math of projectiles, there is an important assumption you need to make to simplify the problem. The basketball in your problem will not be traveling so fast, nor for so long, that air resistance will have a large effect, so you will ignore it in this problem. This then means that gravity is the only force affecting the ball after it has been thrown.

In this kind of problem, it really helps to split the motion of the ball into two parts, or components—horizontal and vertical. As you know that gravity is the only force that affects the ball, and gravity acts vertically downwards, then the horizontal speed of the ball must be constant. So, you have a trade-off between vertical height and horizontal range for your trajectory. If you throw the ball at too steep an angle, it won't have the horizontal movement you need to reach the hoop, as too much of the launch speed will go into beating gravity to make the ball go high. If you throw the ball at too low an angle, it will fall below the hoop before it reaches it, as not enough of the launch speed has gone into beating gravity.

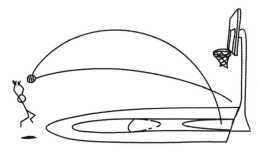

It makes sense, then, that the best compromise is in the middle. To get the ball to go the farthest for a given launch speed, you need to share the horizontal and vertical speeds equally, which happens if you launch the ball at 45° to the horizontal. It also allows you to leave trigonometry out of the solution, as you can use Pythagoras's theorem (see Introduction) instead to find the launch speed you need. We'll assume you throw the ball from the level of the hoop, which makes the mathematics simpler, too.

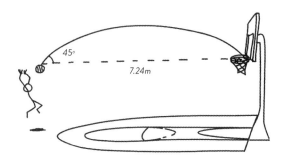

To relate launch velocity (V) to the horizontal and vertical velocity components that I've called u, you can use Pythagoras. If the ball goes V meters on the diagonal, then the horizontal and vertical components must be equal, as the triangle is an isosceles. This means that, in this context, Pythagoras's theorem would look like this:

$$V = \sqrt{u^2 + u^2}$$

$u^2 + u^2$ is $2u^2$:

$$V = \sqrt{2u^2}$$

Using the same logic that you used in relation to the dinosaur bone on page 59,

$$V = \sqrt{2}u$$

The equation that will sort the rest of this problem out for you is one of Newton's equations of motion, and it looks like this:

$$s = ut + \tfrac{1}{2}at^2$$

You're going to look at two versions of this equation. One is for the horizontal motion of the ball, and the other for the vertical motion. Here s is for displacement, which is the distance of the ball from where it was thrown. The component of launch speed—either horizontal or vertical—is u, t is for time and a is for acceleration.

Horizontally, there is no acceleration, which means a = 0, so you can lose the whole $\frac{1}{2}at^2$ part:

$$s_h = ut$$

The h subscript is to remind us that this is for the horizontal direction. Vertically, the acceleration is due to gravity. This distance traveled vertically is defined as being positive as the ball travels upwards. Gravity acts in the opposite direction to this, so it is negative in this equation. You substitute in a = -g:

$$s_v = ut + \frac{1}{2} \times -g \times t^2$$

Tidying up the right-hand side of the equation gives

$$s_v = ut - \frac{1}{2}gt^2$$

As you want to find out at what speed you need to throw the ball, you want an equation that says "u =" and you don't want t to be involved, as you have no idea how long the shot will take and don't want to have to work it out. You can eliminate t by rearranging the first equation to show t in terms of s_h and u:

$$s_h = ut$$

Record breakers

In 2014, Elan Buller from the US made a basket from 34.3 m away. This makes his launch speed a phenomenal 18.3 m/sec, nearly 66 km/h. Quite a throw. The highest basketball shot ever is just over 200 m, thrown from the top of a waterfall in Lesotho. Air resistance plays a huge role in such a shot. As the ball spins, it begins to generate lift, which increases with speed. In much the same way as a proficient soccer player can curl the ball, Australian Derek Herron had to take this effect into account when he made his shot, although it did take six days to get it right.

Dividing both sides by u gives

$$\frac{s_h}{u} = t$$

Here, t—time—is the same in both the horizontal and the vertical equation. This means you can substitute the result above into the vertical motion equation

$$s_y = ut - \tfrac{1}{2}gt^2$$

With the substitution, it becomes

$$s_y = u\,\frac{s_h}{u} - \frac{1}{2}\,g\left(\frac{s_h}{u}\right)^2$$

This may look daunting, but it can be simplified a bit as u ÷ u = 1 and you can square the bracket:

$$s_y = s_h - \frac{g{s_h}^2}{2u^2}$$

You can then take it step by step to find u:

$$s_y = s_h - \frac{gs_h^2}{2u^2}$$

Subtracting s_h from both sides gives

$$s_y - s_h = \frac{gs_h^2}{2u^2}$$

Multiplying both sides by u^2,

$$u^2(s_y - s_h) = -\frac{gs_h^2}{2}$$

Then, you can divide by the bracket on the left-hand side

$$u^2 = -\frac{gs_h^2}{2(s_y - s_h)}$$

You take the square root of both sides,

$$u = \sqrt{-\frac{gsh^2}{2(s_y - s_h)}}$$

Finally, we can put some numbers in: g is 9.81 m/sec², s_h is the 7.24 m to the hoop horizontally, and s_y is zero, as we want the ball to be vertically level with the place where we launched the ball.

$$u = \sqrt{-\frac{9.81 \times 7.24^2}{2(0-7.24)}}$$

$$u = 5.9592 \text{ m/sec}$$

Now that you know the horizontal and vertical components of the launch speed, you can use them to work out the launch speed itself:

$$V = \sqrt{2}u$$

Substituting in u = 5.9592,

$$V = \sqrt{2} \times 5.9592$$

$$V = 8.43 \text{ m/sec}$$

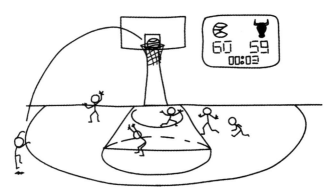

If you throw the ball faster than this, you get two possible trajectories that would allow you to score. With the extra velocity, you can throw the ball at a higher angle, perhaps above the opposing team's giant star player. Or, you can throw a faster, flatter trajectory, maybe to avoid giving the opposition time to block your shot.

Having performed all these ballistic calculations in your head in mere moments, you throw the ball at 45° and at 8.43 m/sec and win the championship for your team. Basketball scholarships, an NBA career and universal adulation are yours for the taking!

CHAPTER 10

A Brisk Walk

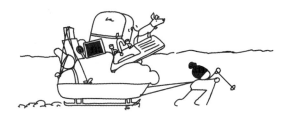

Walking unaided to the South Pole is the challenge of a lifetime. Famously claiming the lives of all five of Robert Scott's team in 1912, the terrain, the cold and the isolation all contribute to making any attempt a potential life-or-death struggle. As a world-renowned explorer and ultra-endurance specialist, you are keen to face this ultimate test, as well as to use it to raise a huge amount of money for your charities. You intend to drag a sled carrying all your equipment and rations to see you through the 120-day, 2,900-km journey. You have all the latest lightweight equipment, but you need to calculate exactly how much food you will need to bring with you in order to survive. The more you bring, the heavier the sled, so the more calories you burn pulling it, the more food you need. Don't bring enough, however, and you won't have the energy to complete the trip. Due to both environmental concerns and medical research, you

won't be able to leave anything behind, including any human waste products, and you have been asked to collect some ice samples as you travel, so the sled will not get lighter.

Calories. We've all heard this word, and many of us make our food choices based on them. But what is a calorie? First, a calorie is not a calorie. A calorie, as far as food is concerned, is actually a kilocalorie, or 1,000 calories. A kilocalorie (kcal) is the energy required to increase the temperature of a kilogram of water by one degree Celsius. Scientists usually use the Joule (J) as the unit of energy, and a kilocalorie is equivalent to 4,184 J.

Our bodies burn energy all the time, even when we're not moving. The energy is used for essential metabolic processes like pumping blood, breathing, maintaining your body temperature, sometimes even for thinking. It varies with weight, height and build, but an average-sized adult uses approximately 100 kcal/h, which equates to roughly 2,400 per day.

If you use more calories than you consume, your body turns to its reserves—fat, initially, and then muscle. There were many factors in Captain Scott's downfall, but one of them would certainly have been the fact that even though his rations contained roughly 4,300 kcal per day, it was nowhere near enough to cope with the requirements for the hostile conditions and huge exertions involved—and each man lost over 30 kg before the end of their journey.

The main energy food groups are fat, carbohydrates and proteins. Carbohydrates and proteins each provide about 4 kcal/g, but fat has more than double this at 8.8 kcal/g. This is one of the reasons our bodies store excess food as fat, and also one of the reasons it can be very difficult to lose weight. One kilogram of fat represents 8,800 kcal—more than a day in the saddle for a professional cyclist—and the equivalent of over 3.5 days of normal calorie intake. Fiber—essential for efficient digestion of all these calories—comes in at 1.9 kcal/g.

It makes sense to skew your diet towards fat for the Antarctic marathon before you, but you need to work out how much energy you will get from your diet. By mass, you and your nutritionist create rations that are 50 percent fat, 20 percent protein, 20 percent carbs and 10 percent fiber. If we consider the average contents of 10 kg of your rations,

Food group	Mass in grams	Energy calculation	Energy in kcals
Fat (8.8 kcal/g)	5,000	8.8 × 5,000 =	44,000
Protein (4 kcal/g)	2,000	4 × 2,000 =	8,000
Carbohydrate (4 kcal/g)	2,000	4 × 2,000 =	8,000
Fiber (1.9 kcal/g)	1,000	1.9 × 1,000 =	1,900

This gives us a total of 44,000 + 8,000 + 8,000 + 1,900 = 61,900 kcal in 10 kg, so 6,100 kcal/kg.

You will use this energy in two ways. First, you need to ensure that you have the daily survival amount of 2,400 kcal. Of course, in addition to this, you'll use a lot of extra energy dragging yourself and the sled across the untracked vastness of Antarctica. If we look at that as an equation, it'd be something like this:

Total energy = dragging energy + survival energy

So, let's first look at how much dragging energy you'll be using, pulling your sled. To get a broad idea, you need to make some initial simplifications. Let's assume that you travel at a constant speed and that air resistance is not a factor. And to start with, we'll also assume the terrain on your journey is flat.

The main force you have to overcome is friction. Friction is a force that works in the opposite direction to the way that you are moving. If you try to pull the sled in one direction, the friction force acts in the opposite. When you are traveling at constant speed, the friction force and the force you are producing must be equal—if not, you would either speed up or slow down.

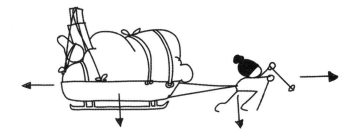

Friction depends on two factors. The first is the materials that are in contact with each other—in our case the runners on the bottom of the sled and the snow or ice itself. How well a sled moves across the Antarctic surface really depends on the exact nature of the surface itself. Just as the Inuit have many words for snow, each type of snow will produce more or less friction with the sled.

The second factor is the mass of the sled. The heavier it is, the more friction you will need to overcome. The relationship is given by this formula for a moving object:

$$F = \mu W$$

The symbol μ (the Greek letter "mu") is called the coefficient of friction and represents the relative friction between the two surfaces. As mentioned, it varies a lot for snow, but you'll use an averaged figure of 0.2 in your work. W represents weight—in this case the weight of you and the sled combined. Now, the mass of the sled consists of the mass of the structure of the sled itself, as well as all the cargo and food. The sled has a mass of 25 kg, your equipment such as tents, fuel, stoves, etc. comes to another 100 kg, plus the mass of food, which we'll call m_f. This makes the mass of the loaded sled equal to $25 + 100 + m_f = 125 + m_f$. We'll say your mass is 70 kg, giving a total mass of

$$\text{Total mass} = 125 + m_f + 70$$

$$\text{Total mass} = 195 + m_f$$

Weight is mass times g, so the total weight is $(195 + m_f)g$. Putting this into the friction formula gives you

$$F = \mu(195 + m_f)g$$

If you substitute in for μ and g, you can work out the friction force you need to overcome in order to move across Antarctica,

$$F = 0.2 \times (195 + m_f) \times 9.81$$

This can be rearranged as

$$F = 0.2 \times 9.81 \times (195 + m_f)$$

Multiplying 0.2×9.81 gives

$$F = 1.962 \, (195 + m_f)$$

Expanding the bracket gives

$$F = 1.962 \times 195 + 1.962 \times m_f$$

$$F = 382.59 + 1.962 m_f$$

So, you might be wondering, how do you relate this force with all the previous talk about energy? Well, the energy or work done in producing a force is simply the force multiplied by the distance, d:

$$\text{Dragging energy} = \text{friction force} \times \text{distance}$$

$$\text{Dragging energy} = (382.59 + 1.962m_f) \times d$$

The distance you have to walk, d, is 2,900 km, which is 2,900,000 m:

$$\text{Dragging energy} = (382.59 + 1.962m_f) \times 2{,}900{,}000$$

Multiplying out the brackets gives

$$\text{Dragging energy} = 382.59 \times 2{,}900{,}000 + 1.962m_f \times 2{,}900{,}000$$

$$\text{Dragging energy} = 1{,}109{,}511{,}000 + 5{,}689{,}800m_f$$

This energy is in Joules, the standard unit of energy. You realize it will be easier to work in kilocalories, so you can convert this by dividing by 4,184:

$$\text{Dragging energy} = 265{,}179 + 1{,}360m_f$$

Now, you move on to the survival energy requirements. You know that you need 2,400 kcal per day to avoid keeling over in a sorry heap in the snow, and you also know that the trip will take 120 days, giving

$$\text{Survival energy} = 120 \times 2{,}400 = 288{,}000$$

So, you need a whopping 288,000 kcal just to stay alive on the trip. Your life-saving energy equation now looks like this:

$$\text{Total energy} = \text{dragging energy} + \text{survival energy}$$

$$\text{Total energy} = 265{,}179 + 1{,}360m_f + 288{,}000$$

$$\text{Total energy} = 553{,}179 + 1{,}360m_f$$

All the energy required is provided by the food. You know that each kilogram of food will provide 6,100 kcal of energy, so the total energy must be the mass of the food multiplied by the 6,190 kcal you get from each kilogram:

$$6{,}190m_f = 553{,}179 + 1{,}360m_f$$

To solve this equation, you can gather all the m_f terms together by subtracting $1{,}360m_f$ from each side:

$$6{,}190m_f - 1{,}360m_f = 553{,}179$$

$$4{,}740m_f = 553{,}179$$

Finally, you divide both sides by 4,740 to find out how much food you'll need:

$$mf = 553{,}179 \div 4{,}740$$

$$mf = 117 \text{ kg (nearest kg)}$$

You can see that this is just under 1 kg of food, 6,190 kcal, per day. But in reality, of course, you will need a bit more. Your model stated that you will be moving at a constant speed. But again, in reality, you won't. There will be a lot of stopping and starting, taking rests, checking GPS, having a snack, answering the call of nature. Every time the sled slows, it takes energy to get it moving again, as you have to provide enough force to accelerate the sled as well as overcome the friction.

Your model also assumed that you would be traveling on a level surface. If you start your trip on the Ross Ice Shelf like Scott and his ill-fated team, you'll be somewhere between 15 and 50 m above sea level. The South Pole itself has an altitude of 2,835 m, and between the ice shelf and the pole are the Trans-Antarctic mountains, with peaks as high as 4,500 m. You will definitely have some climbing to do! Ignoring air resistance may be optimistic, too. The weather in Antarctica is extreme. Even in the summer, the average temperature is -26°C, and blizzards are possible at any time. Cold air falling down from the altitude of the South Pole produces a phenomenon called a katabatic wind, which can exceed hurricane force. Not something you want to walk into. You pack your sled and decide to add some extra rations after all.

CHAPTER 11

The Parachute Equation

You are enjoying a small glass of champagne in the business compartment of the airliner taking you home after a very successful business trip. You have struck up an interesting conversation with your neighbor, Avantika, who is an international business traveler like yourself, exploiting the current trend in saris to extend her family business to VIP clientele across the world. Suddenly, the plane rocks, and you lurch forward against your seatbelt as loud bangs come from the engines and the front of the plane. Ignoring the ice-cold beverage now in your lap, you watch as a flight attendant tries and fails to get a response from the flight deck. "Bird strike!" you see him mouth to another member of the cabin crew. It's then that you notice that the ever-present whine of the engines has ceased, and you feel the plane start to descend. As you glide over the moonlit jungles of Southeast Asia with no engines, you then realize that the life jacket under your seat is pointless, and only a parachute will save your life. And then you think of all the yardage of silk that you know your sari-selling friend has in her carry-on. Is it possible to fashion a working parachute before the plane crashes?

Modern airliners are marvelous things. According to research by Northwestern University, there are 0.07 deaths per billion passenger miles on aircraft, whereas there are 7.28 deaths per billion miles in cars and a whopping 213 on motorcycles. Statistically speaking, you are over a hundred times more likely

to die in a car than an aircraft. Planes, however, travel very quickly—500 knots or about 900 km/h. So, when they hit things—such as a flock of geese—they hit hard. Exactly this happened to a US Airways flight from New York in 2009. Fortunately, the plane was expertly ditched in the Hudson River, and everyone survived. Go watch the film *Sully: Miracle on the Hudson* for more detail.

People have had the concept of parachutes since at least Leonardo da Vinci, who sketched a decent one in 1485. This was successfully created and used by British skydiver Adrian Nicholas in June 2000.

The underlying principle of a parachute is simple. When an object falls through air (or any other gas or liquid for that matter), it experiences a drag force. The drag force increases as the object's speed increases, until it equals the gravity force pulling it downwards, known as weight. At this point, the object stops

Drag

Weight

accelerating. The speed reached depends on the size and shape of the falling object, and is known as terminal velocity. For a person in the "belly-to-earth" skydiver position, this is about 55 meters per second or 120 miles per hour (mph), and it would be fatal if you hit the ground at this speed. The parachute massively increases the surface area of the falling object, meaning it must push much more air out of the way as it falls. Thus, the drag force is much bigger and the terminal velocity is much lower. With a modern parachute, the figure is much more survivable: 5 m/sec or 11 mph.

Who needs a parachute anyway?

A few people have survived falling out of planes without parachutes, most of them sustaining, though, some pretty serious injuries. A couple of people were even luckier. During the Second World War, British Flight Sergeant Nicholas Alkemade's Lancaster was attacked after a raid on Berlin and spun out of control. His parachute had been burned during the fire, but Alkemade decided to jump anyway, preferring death from falling to burning. He fell from five or six kilometers in altitude, but, miraculously, had his fall broken by pine trees and landed in a deep snowdrift. He was able to walk away with only minor injuries. And in 2012, stuntman Gary Connery became perhaps the first person to survive an intentional jump from an aircraft with no parachute. He dropped 730 m from a helicopter near Henley-on-Thames and used a wingsuit to steer himself to a landing in eighteen thousand cardboard boxes.

So, we want to balance weight and the drag force. Weight is given by an object's mass multiplied by the acceleration due to gravity. The drag force is related to the density of the air, the speed of travel, and the size and shape of the object:

$$\text{Weight} = \text{drag force}$$

$$mg = \frac{1}{2}\, \rho C_d A v^2$$

Don't be alarmed by all those letters on the right-hand side. A is the cross-sectional area of the parachute—the area of the circle at the mouth of the parachute; m is the mass of the falling object and parachute; g is the acceleration due to gravity; the Greek letter "rho," ρ, is the density of the air; C_d is the drag coefficient of the parachute—a measure of how aerodynamic the parachute is; v is the descent speed desired.

As you are going to have to make the parachute and need to know how big to make it, you rearrange the equation to make $\overset{\text{\tiny .}}{A}$ the subject. First, you divide both sides of the equation by $\rho C_d v^2$:

$$\frac{mg}{\rho C d v^2} = \frac{1}{2}\, A$$

Then, you can multiply both sides by two to give a formula for A:

$$A = \frac{2mg}{\rho C d v^2}$$

So, to put some numbers into the equation:

m is your mass in kilograms, plus that of the parachute—
let's say 100 kg.

g is 9.81 m/sec². On Earth, at least.

ρ is 1.2 kg/m³ at sea level, and working on the principle that you are going to jump out of the plane as late as possible, this is reasonable.

C_d is tricky. It is very difficult to work out mathematically what the drag coefficient of any particular shape is, so we tend to evaluate this experimentally using wind tunnels or just timing things falling. A nice dome-shaped parachute has a drag coefficient of about 1.5, whereas a flat surface has a drag coefficient of about 0.75. Your sari parachute is probably going to be somewhere between the two, so you estimate it at about 1.1.

v you can choose to some extent, and the larger you make it the smaller the parachute needs to be. Hitting the ground at 9 m/sec or 20 mph is the equivalent of falling about four meters. Not nice, but survivable and hopefully without injury with a bit of luck.

Therefore,

$$A = \frac{2 \times 100 \times 9.81}{1.2 \times 1.1 \times 9^2}$$
$$= 18.4 \text{ m}^2$$

This is the area you need at the mouth of the parachute. The area of a circle is given by πr^2, so,

$$18.4 = \pi r^2$$

You need to rearrange this to make r the subject, which you do by dividing both sides by π and then square rooting:

$$r = \sqrt{\frac{18.4}{\pi}}$$

$$r = 2.42 \text{ m}$$

So, to make the parachute, you need to make a hemisphere (half-sphere) shape out of the sari material. To do that, you need to know the amount of sari this will require. The area of the surface of a hemisphere is given by $2\pi r^2$. This happens to be twice the circular cross-sectional area, which is πr^2. If you need an 18.4 m² cross-sectional area, you need $18.4 \times 2 = 36.8$ m² of sari parachute.

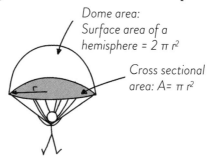

Dome area:
Surface area of a
hemisphere = $2\pi r^2$

Cross sectional
area: $A = \pi r^2$

The length of the curved surface of the canopy is half the circumference of the sphere. As circumference = $2\pi r$, so half of that is πr.

$$\text{Length of curved surface} = \pi r$$
$$= \pi \times 2.42$$
$$= 7.60 \text{ m}$$

You turn to Avantika and blurt out your idea to make a parachute using her saris. She thinks you are mad but is happy to help to keep her mind off the impending crash-landing. She tells you that her sari cloths are 8 m long and 1 m wide, which means that five of them will give you 40 m² of material and be about the right length to give the correct cross-section. But is there time to sew them together?

When anything starts to glide in still air, we need to consider three forces: lift, drag and weight.

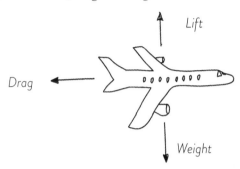

Without a source of power such as an engine, updrafts or thermals, it isn't possible for a glider to climb without sacrificing speed. This means that the plane will descend as it moves forward. How much it descends depends on something called the glide ratio, which itself depends on how fast the plane is going. The glide ratio tells you how far forward you travel for every meter the glider loses in height. For an airliner, this is usually in the region of 15 to 20:1, meaning that you travel 15 to 20 m forward for every meter descended.

Let's assume your plane has a glide ratio of 17.5:1. If your plane was at about 8,000 m in altitude (well within the capabilities of many migratory birds, by the way), then it would travel 8,000 × 17.5 = 140 km before reaching sea level. This seems a very long way, but remember that airplanes travel quickly. Airliners tend to have their best glide ratio at about 250 knots, which is roughly 130 m/sec. To work out how long you have, you need to use the fact that time is given by distance divided by speed:

$$\text{Time} = \frac{\text{distance}}{\text{speed}}$$

The distance in meters is 140,000 at a speed of 130 m/sec:

$$\text{Time} = \frac{140,000}{130}$$

$$\text{Time} = 1,077 \text{ sec}$$

This is just under eighteen minutes. Enough time to fashion a parachute? Let's hope so! But how to make the parachute the right shape?

Most dome parachutes are made using triangular-shaped pieces of ripstop nylon, stitched together to make a large circle. You don't have anything like enough time to do that. Fortunately, Avantika does have some miracle silk glue that will stick the saris together quickly and strongly. How can you make a roughly circular parachute with rectangular strips of silk?

Perhaps the quickest way that does not require gluing long seams together would be to overlap the strips in the middle, to make an asterisk shape:

You see that you will end up with a ten-pointed star shape. This may be quick to make, but the overlap is going to cost you some surface area. Avantika suggests using six saris to compensate, but would this still be enough? What area are you left with?

It turns out that using six saris, forming a twelve-pointed shape, makes the calculations a bit easier, so you stick with that. We can work out the total area by looking at one-twelfth of the shape:

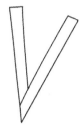

Luckily, with her mastery of textiles and pattern cutting, Avantika can tell you that this gives an area 3.07 m². You multiply this by twelve to get the total area. As if by magic, you get a result of 36.8 m², almost exactly what you need. Someone up there still likes you.

You salvage some cord from the life jackets around the cabin and lash them to yourself and the ends of the parachute as securely as you can. The plane has descended enough that you can breathe without your oxygen mask, so you decide now is as good a time as any to make the jump. You give Avantika a hug and say goodbye, and she shakes her head in disbelief that you will trust your life to the makeshift armful of silk and lifebelt that you hold in your arms. A steward helps you to a small hatch in the underbelly of the plane. You start a countdown …

Knots and nautical miles

For day-to-day travel, we use kilometers and miles to measure how far we need to go, which assume you are traveling on a flat surface. If you are traveling farther, you need to take into account that, despite what the Flat Earthers say, the Earth is a sphere, and so you are in fact traveling along a curve. A nautical mile is defined as one-sixtieth of a degree of latitude and works out at 1.852 km or 1.15 miles. A knot is a speed of one nautical mile per hour, and stems from the days when sailors would measure the speed of their vessel by throwing a panel of wood over the stern of their ship and counting how many knots on the rope towing the panel passed through their fingers in half a minute. The knots were eight fathoms (a fathom is six feet) apart to give an approximate speed. Aviators also measure their speed in knots for similar reasons to mariners, but I have left the calculations in this chapter in the more familiar kilometers per hour for simplicity.

CHAPTER 12

Space Rescue

As a pilot of the next generation of space shuttle, you regularly ferry supplies, equipment and space tourists to the new space resorts run by private consortiums and technology billionaires. You are preparing to depart one station, Spacebados, when a distress call comes from another franchise, No Geeland. A meteorite strike has left them in trouble and they require evacuation. Can you get from your orbit to theirs and save the day? You are currently in orbit in your 150-tonne shuttle, 400 km above the Earth, and need to rendezvous with the other station 50 km farther out. Can you reach them?

We've all seen footage of rockets launching. Huge white monolithic machines, shuddering with barely controlled power, with the strength and endurance to leave the Earth and go into space itself. We see them return, charred and bruised by their unforgiving voyage, and—tragically—we've seen catastrophic and fatal outcomes when things go wrong. Traveling to space is the ultimate challenge for humans and machines alike.

Or is it?

Space is 100 km above our heads. That's not really very far. Nazi V2 rockets, first built in 1944, became the first artificial objects to enter space, and even amateur rocket enthusiasts managed to send a rocket into space in 2004. So, getting to space has been possible for some time, and you clearly don't need to work for NASA to make it happen. But, it's staying there that is the hard part.

As we've seen in previous chapters, gravity varies with distance from the Earth. At 100 km, acceleration due to gravity is about 9.8 m/sec². This is 97 percent of the gravity at the Earth's surface. Essentially, we can fire things straight up for 100 km and they will be in space, but then they will fall back down again. Yet we know the Earth is surrounded by satellites of all shapes and sizes: telecom satellites, sinister government spy sats and space lasers, the International Space Station and even the moon. What stops them from falling into the Earth? What makes them stay there? Why do astronauts on board space stations seem to be weightless?

To understand why this is, we can go through a very similar thought process to the one Newton (yep, him again) did when he first worked this out. If you throw a ball horizontally, it follows a curved path and hits the ground. The faster you throw it, the farther it will go. As the ball travels, the Earth curves away underneath it, which is not normally an issue for most throws as they don't go far enough for the curvature of the Earth to have an effect. Again, Newton's genius was to see past this and ask: what if you threw it really, really hard?

Ignoring tedious things like air resistance and objects and mountains being in the way, if you threw the ball hard enough, it would follow the curve of the Earth and go all the way around the planet and hit you in the back of the head. Or, if you moved out of the way, it would carry on going around the Earth. In this case, the ball is said to be in orbit around the Earth. You'd have to throw the ball at exactly the right speed to make it do this—too slow and it would hit the ground, too fast and it would fly off into space.

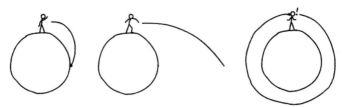

The ball would be traveling in a circle. When you are in a vehicle that is going around a bend (i.e., a small part of a circle) you feel a force pulling you sideways. This is because you are changing direction, which can only happen due to a force. When a car takes a corner, friction from the wheels provides the force going into the corner. There is a limit to this friction, so if you take a corner too quickly the car will skid outwards, a bit like throwing the ball too hard opposite. We know from experience that the force we feel when moving in a circular path depends on how fast we are going and also how tight the curve is (i.e., the radius of the curve).

The mass of the object moving also matters. Think of having an object on a string that you are whirling around you, lasso style. The heavier that object, the more effort you have to put into the whirling. The relationship is shown by this formula for the force in circular motion:

$$F = \frac{mv^2}{r}$$

Anything following a circular path needs to have this force pulling it into the center of the circle or it will not follow that path. In the lasso example, this force is tension in the rope. Cut the rope, and the object will fly off, out of the circle.

Satellites (and balls thrown very hard) are not attached by ropes, so their circular motion force must be provided by gravity—their weight provides the force.

So, the ball you threw follows that curved path around the planet because of gravity constantly pulling it downwards—in other words, falling, albeit while moving along however fast you threw it. This is why astronauts on space stations appear not to have gravity. They are, in fact, in free fall, constantly falling but always missing the Earth.

If you really wanted to hit yourself with the ball, you'd be on the surface of the Earth, so you could use W = mg.

$$mg = \frac{mv^2}{r}$$

You have mass on both sides of this equation, so you can divide both sides by mass to eliminate it. This tells you that the mass of the object in orbit does not affect the speed required. You can now rearrange the formula to make v the subject so that you can work out how fast you need to throw the ball to get it to orbit the Earth:

$$g = \frac{v^2}{r}$$

To isolate v, the next step is to multiply both sides by r:

$$gr = v^2$$

Then you take the square root of each side:

$$\sqrt{gr} = v$$

The radius of the Earth is 6,371 km and g is 9.81 m/sec², which gives you:

$$v = \sqrt{9.81 \times 6,371,000}$$

$$v = 7,906 \text{ m/sec}$$

That's a fairly nippy throw, clocking in at about 28,000 km/h. For context, the fastest throws from professional baseball pitchers come in at about 160 km/h. The fastest arrow shot by hand comes in at over 600 km/h. The fastest jet aircraft in the world, the iconic Lockheed SR-71 Blackbird, can only manage a measly 3,500 km/h. If you take into account air resistance, there is no way you could launch anything at the required speed without its burning up due to friction with the atmosphere pretty quickly.

Throwing balls at ground level is all well and good, but your shuttle is at Spacebados, which orbits at a height of about 400 km (6,771 km above the center of the Earth). So, rather than having mg on the left-hand side of the formula, you need to use the gravitation formula to take into account the reduction in gravity. As there are now two different masses to take into account in the equation, we'll call the mass of the Earth m_E to avoid confusion:

$$\frac{Gm_Em}{r^2} = \frac{mv^2}{r}$$

Again, I can cancel the m—the mass of the space station—on each side of the equation as well as an r on each side:

$$\frac{Gm_E}{r} = v^2$$

Taking the square root of each side gives me a formula for v:

$$\sqrt{\frac{Gm_E}{r}} = v$$

We know the values of G (0.0000000000667 m³/kgs²) and m_E (5,970,000,000,000,000,000,000,000 kg) from previous chapters. The radius is 6,371 km plus the 400 km above the Earth, giving 6,771 km. Putting these in gives the speed of your shuttle, docked to the station, as:

$$v = \sqrt{\frac{0.0000000000667 \times 5,970,000,000,000,000,000,000,000}{6,771,000}}$$

$$v = 7,671 \text{ m/sec}$$

So, to get your space shuttle to Spacebados in the first place, you needed enough energy to raise the rocket 400 km upwards against gravity and enough to give it a speed of 7.7 km every second. This is a bit slower than the ball. This is because r is on the bottom of the fraction in the formula, meaning that as r gets bigger, the speed required will get smaller as you are dividing by a bigger number. The higher the orbit, the lower the speed you need to maintain that orbit. Does that mean you need to slow down to reach the higher orbit and perform the rescue, I hear you ask?

This is where you need to consider the energy required for the orbit. We are interested in two kinds: the energy the shuttle has from moving, called kinetic energy, and the energy the shuttle has in the form of work done against gravity, called gravitational potential energy. To help you wrap your head around the latter, imagine a family heirloom, perhaps a delicate vase, sitting on a shelf. It appears to have little or no energy—after all, it is not moving. But when your cat jumps up onto that shelf and tips it off, it accelerates rapidly towards the ground and shatters. Where did this energy come from? It came from whoever lifted it up there in the first place. Things have the potential to do things like fall or roll downhill simply from being higher up, so they must have energy.

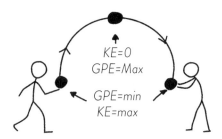

So, the kinetic and gravitational energy (KE and GPE for short) are in a trade-off. Imagine throwing a ball into the air. You give it KE when you throw it, which gradually turns into GPE as it gains height. When all the KE has turned into GPE, the ball has reached its apex. The GPE then turns back into KE as it falls back down again. At any given point, the ball has the energy you gave it by throwing it. This energy gets shared by the ball's KE and GPE as it travels, so we could write

$$\text{Total energy} = \text{KE} + \text{GPE}$$

The same holds true for your shuttle. In a given orbit, burning fuel has provided a certain amount of energy, which is shared between KE and GPE. We need to look at how much energy we have in our current orbit, versus how much we require for the new orbit. The formulae for the energies are, with m_s as the mass of the shuttle and m_E as the mass of the Earth,

$$KE = \frac{1}{2} m_s v^2$$

$$GPE = - \frac{Gm_E m_s}{r}$$

There is something slightly peculiar about GPE though—it is defined as being negative. Why? When an object is really far away from Earth (and other heavy things), gravity won't affect it. If you let it go, it wouldn't fall. If the object has zero energy far away, it must have negative energy as we get closer to the Earth, so that the amount of GPE still increases as we go up, but we are starting from a negative standpoint. It is similar to depth in the ocean—sea level is zero, and anything under the ocean is negative altitude. Scientists actually refer to the zone of influence of a planet's gravity as its gravity well.

The formulae opposite are similar to those for the forces, and there's a good reason for this. We saw in Chapter 9 that energy can be calculated by multiplying a force by a distance. There's a little more to it than that for things with changing velocities and gravitational forces, but the reasoning is the same.

Now, you've worked out opposite the velocity required for a certain orbit. If you use this in your KE formula,

$$KE = \frac{1}{2} m_s v^2 \text{ where } v = \sqrt{\frac{Gm_E}{r}}$$

You want v^2, and squaring v will cancel out the square root symbol:

$$KE = \frac{1}{2} \, m_s \times \frac{Gm_E}{r}$$

You can combine this into one fraction:

$$KE = \frac{Gm_s m_E}{2r}$$

This looks very similar to the GPE formula. The total energy must then be:

$$\text{Total energy} = \frac{Gm_s m_E}{2r} - \frac{Gm_s m_E}{r}$$

This is similar to taking one away from a half—you get minus a half:

$$\text{Total energy} = - \frac{Gm_s m_E}{2r}$$

If you can compare the energy for this orbit and the energy for the new orbit, you can tell whether you need to add or subtract energy (in the form of fuel in the engines) to the spacecraft. Negative energy would mean slowing the rocket down, positive would mean speeding it up. The difference in energy is the energy of the new orbit with radius r_2, minus the energy of the original orbit with radius r_1.

$$\text{Change in energy} = - \frac{Gm_s m_E}{2r_2} - - \frac{Gm_s m_E}{2r_1}$$

Remembering that subtracting a negative amount is the same as adding the amount,

$$\text{Change in energy} = - \frac{Gm_s m_E}{2r_2} + \frac{Gm_s m_E}{2r_1}$$

Which is the same as

$$\text{Change in energy} = \frac{Gm_sm_E}{2r_1} + \frac{Gm_sm_E}{2r_2}$$

You notice that the numerators of each fraction are the same, and they have both been divided by two, so for ease of working you calculate their value. You know G and m_E from before, and we are told that m_s is 150 tonnes, which is 150,000 kg:

$$\frac{Gm_sm_E}{2} = \frac{0.0000000000667 \times 150,000 \times 5,970,000,000,000,000,000,000,000}{2}$$

This gives:

$$\frac{Gm_sm_E}{2} = 29,880,000,000,000,000,000$$

You put this back into the change in energy equation:

$$\text{Change in energy} = \frac{29,880,000,000,000,000,000}{r_1} - \frac{29,880,000,000,000,000,000}{r_2}$$

You remember that the radius of the Earth is 6,371 km, making the lower orbit 6,371 + 400 = 6,771 km, and the higher one is 50 km more than this at 6,821 km:

$$\text{Change in energy} =$$

$$\frac{29,880,000,000,000,000,000}{6,771,000} - \frac{29,880,000,000,000,000,000}{6,821,000}$$

$$\text{Change in energy} = 32,300,000,000 \text{ J}$$

Going beyond

Traveling in space is complicated. Changes in gravity, changes in mass as your craft uses up fuel and changes in position of your destination make plotting a course a serious challenge. One thing that is for certain is that you cannot simply aim your craft at your destination and light the rockets. In 1977, two Voyager space probes were launched to visit the planets Jupiter, Saturn, Uranus and Neptune, which were in a rare grouping that made visiting them all possible. Following trajectories carefully calculated to take them near each planet and using the planet's gravity to bend their trajectory towards the next planet, it took Voyager 1 eighteen months to reach Jupiter and a further twenty months to reach Saturn. It took Voyager 2 twelve years to reach Neptune. Since the late '80s, both probes have continued traveling. Voyager 1, at the time of writing, is the farthest human-made object from the Earth, over 22 billion km away in interstellar space after a forty-two-year journey. Every second sees it travel another 17 km.

The figure of 32.3 billion J is most definitely a positive number, so in order to get higher, you need to add energy to the system by burning fuel to speed up. It may seem counterintuitive, but to get to a higher, orbit in which you are going slower, you need to accelerate. Although this may seem like a huge amount of energy, it corresponds to burning about 250 kg of hydrogen with 125 kg of oxygen. This will produce 375 kg of steam as exhaust from the shuttle.

As you burn fuel to accelerate the shuttle, the gravity force is no longer enough to keep you going in the same circular orbit, so the shuttle will drift outwards into an egg-shaped elliptical orbit.

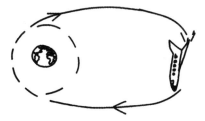

In this orbit, your shuttle slows down as you get farther from the Earth, again exchanging KE for GPE. When you get to the right altitude, another burn of fuel will accelerate the rocket to the correct speed for circular motion, changing the elliptical orbit back into a circular one.

With the help of ground control, you time the burns to bring you into orbit with No Geeland, where you rescue the grateful (and very wealthy) space tourists and (not so wealthy) crew.

To get back to Earth, you have to reduce the energy of your orbit, so you turn the shuttle around and boost back against the direction of travel to get down to terra firma.

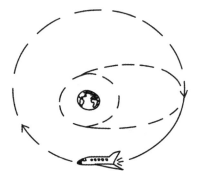

CHAPTER 13

Double or Nothing

Your old Uncle Ebenezer always was a bit of a wild card. He constantly accused you of being a lazy, good-for-nothing spendthrift, and so you're very surprised to receive a letter from his lawyer shortly after his funeral. Even from beyond the grave, the old man teases you. His lawyer says that the terms of his will state that, if you can prove your financial acumen, a rich reward awaits. The letter explains that Uncle Ebenezer will give you $25,000, and if you can turn this into $50,000 through investment within five years, his lawyer will release another $1 million to you. If you fail, the terms of the agreement say you must refund the original $25,000, which will be donated to your least favorite political party. You'd like nothing more than to prove the old coot wrong and take his money, but how can you do it with as little risk as possible? Is it possible with a good old-fashioned savings plan?

Compound interest has been of, well, interest to mathematicians, bankers, economists and anyone borrowing or lending money since money was invented around five thousand years ago. For a long time, charging interest or usury was considered a sin and was often illegal, too. Recent cases of payday loan companies charging enormous amounts of interest have prompted increased regulation.

We know for sure that compound interest makes savings grow more quickly. The concept is simple—you invest a certain amount of money, which then accrues interest as a percentage of how much you originally invested. If you leave the interest in your savings, you then earn interest on this interest and so on, meaning your investment grows faster than the interest rate would suggest.

A very important factor in interest is how often the interest on your investment is added. Say you invest the $25,000 in a 5 percent per annum savings account that paid out at the end of every year. To increase any number by 5 percent, you need to multiply it by 1.05. The 1 represents the money you started with, and the 0.05 represents the 5 percent increase in decimal form. By the end of the year, when the interest is paid, you will have 25,000 × 1.05 = $26,250. Nice—you've made a tidy $1,250 through your investment.

A different bank offers an account with the same 5 percent annual rate, but paid in two six-monthly 2.5 percent installments.

The multiplier for a 2.5 percent increase is 1.025, and you'll need to do this twice because of the two lots of six-month installments: 25,000 × 1.025 × 1.025 = $26,265.63 (to the nearest penny). This is $15.63 more than the first account—the interest on your interest.

Let's take it to the extremes. Another account gives the 5 percent annual interest but paid out every day. This makes my daily percentage 0.01369863014 percent, but you receive this 365 times a year. To work this out, you need to multiply the $25,000 by 1.0001369863014, 365 times. You remember that multiplying a number by itself several times is the same as raising it to a power, so you can use this much more convenient way of writing it down:

$$25{,}000 \times 1.0001369863014^{365} = \$26{,}281.69$$

You need to use so many decimal places here because rounding up or down by even a tiny amount has a large effect on the final answer. Sparing you the calculations, an account that gave interest every hour would give you $26,281.77, just 8 cents more. If you went for every second, you would end up with $26,281.78, just one penny more.

You can see that increasing the frequency of interest payments has diminishing returns, but you can also see that you are $31.77 better off when 5 percent interest is calculated each day of the year. The overall effect for the year is as if the interest rate were 5.13 percent rather than 5 percent. Every little bit helps.

It would be handy, at this point, to have a formula for interest calculations. You may have noticed that, to increase the investment by a certain amount, you needed to work out a multiplier. The multiplier is the annual interest rate divided by how many times over the year it gets paid out, with 1 added to make the value increase by this amount. You then repeatedly multiply by that multiplier, according to the number of times the interest is paid out over the year. This gives you a formula where T is the total, I is the initial investment, r is the interest rate and n is the number of times it is paid over the year:

$$T = I \times \left(1 + \frac{r}{n}\right)^n$$

This gives you the total for a year, but you realize that you'll need to invest Uncle Ebenezer's money for more than a year to reach your target. If you repeat the process for a second year, you get the compound interest another n times. If you invest for y years, then the formula includes one tiny change—see whether you can spot it:

$$T = I \times \left(1 + \frac{r}{n}\right)^{ny}$$

So, we now have a formula that you can plug values into and see when the total exceeds $50,000. What would be even better is if you could rearrange the formula to make y the subject. However, y is a power in the equation, and unless you remember it from school, it's not immediately obvious what you have to do to "un-power" something that you're chasing in a formula.

The answer is a concept called logarithms. Older readers may remember a time before electronic calculators when log tables and slide rules exploited this idea to perform calculations. The idea of logarithms is relatively straightforward, but using them can be very confusing, which is why they are a subject reserved for advanced students. So brace yourself!

Consider the powers of 10:

$$10^1 = 10$$
$$10^2 = 100$$
$$10^3 = 1,000$$

And so on. Every time you increase the power by one, you multiply by 10. It is possible to have powers that aren't whole numbers, although they are not easy to calculate without electronic assistance:

$$10^{1.5} = 31.6227766$$
$$10^{1.75} = 56.23413252$$

If you can write decimals as powers, that means you can write any positive number as a power of ten. If you want to find what power of ten gives 75, you would need to consider:

$$10^t = 75$$

You could find this by estimation, using your calculator. You know the value of t must be between 1.75 (as $10^{1.75} = 56.23413252$) and 2 (as $10^2 = 100$). A bit of trial and error tells you that $10^{1.875} = 74.99$ to two decimal places. You don't want to have to muck about with guesswork though, and neither have people wanted to throughout the ages, especially mathematicians who needed to solve equations where the unknown was a power.

The answer was devised by Scotsman, John Napier, in the early 1600s. He introduced the idea of logarithms as the inverse of—and therefore a way to undo—raising numbers as powers. He produced the first tables that would allow you to look up values for the powers you need.

If you write $\log_{10}(75)$, this asks the question "What power of 10 is 75?" Back in the day you would then look this up in a book of tables, but now it is all done by your calculator. My calculator says the answer is 1.875061263, which you can check by doing $10^{1.875061263}$, which is indeed 75. The base of the logarithm is called 10. You can specify a different base if it is helpful. For instance, if you were trying to solve $2^a = 10$, you would need to know what power of 2 is 10, which you can obtain with $\log_2(10)$. This is 3.3 and a bit, which makes sense as 2^3 is 8 and 2^4 is 16, so 10 must be 2 to the power of something between 3 and 4.

Now, back to your problem:

$$T = I \times \left(1 + \frac{r}{n}\right)^{ny}$$

It's easier to get the thing raised to the power (in this case the bracket) by itself if you want to use logarithms. So, dividing both sides by I will do it:

$$T = I \times \left(1 + \frac{r}{n}\right)^{ny}$$

At this point it simplifies things to substitute in some numbers. The total you need is 50,000; the initial sum you have is 25,000; the interest rate is 0.05, and the bank pays interest daily, giving:

$$\frac{50,000}{25,000} = \left(1 + \frac{0.05}{365}\right)^{365y}$$

$$2 = \left(1 + \frac{1}{7,300}\right)^{365y}$$

Now, you need to know what power of one and one seven-thousand-three-hundredths is equal to 2, as this must equal 365y, so,

$$\log_{1\frac{1}{7,300}}(2) = 365y$$

Evaluating the left-hand side on a calculator tells you that

$$5,060.320984 = 365y$$

To find y, you just need to divide both sides by 365:

$$13.86389311 = y$$

Converting this into years and days, you can see that the investment will first hit $50,000 after 13 years and 316 days. This is much too long, so what can you do to save faster? It occurs to you that nothing in Uncle Ebenezer's terms and conditions says you can't contribute your own money. If you saved a bit each month—again taking advantage of compound interest—perhaps that would make up the difference?

You start by working out how much the $25,000 would be worth after five years, using the compound interest formula.

$$T = I \times \left(1 + \frac{r}{n}\right)^{ny}$$

I is your initial $25,000; r is the 5 percent interest rate, which is 0.05 as a decimal; n is 365, as the interest is being paid every day of the year; and y is 5, as you are investing for five years:

$$T = 25,000 \times \left(1 + \frac{0.05}{365}\right)^{365 \times 5}$$

A bit of calculation—

$$T = 25,000 \times 1.0001369861825$$

$$T = 32,100.09$$

This leaves you with $17,899.91 to earn from your monthly saving plan. The mathematics of this is slightly trickier than the previous compound interest formula. Say you saved $100 per month, again at 5 percent annually, paid out monthly, giving a monthly rate of 0.417 percent. Your savings would grow like this:

Month 1: $100

Month 2:   (100 × 1.00417) + 100 = $200.42

Month 3:   (100 × 1.00417^2) + (100 × 1.00417) + 100 = $301.25

Month 4:   (100 × 1.00417^3) + (100 × 1.00417^2) + (100 × 1.00417) + 100 = $402.51

---

e

Jacob Bernoulli was a Swiss mathematician who was interested in compound interest. He discovered that a bank account with 100 per cent yearly interest (nice if you can get it!) will give you 2.71828 times your initial investment after a year. That value—which later was given the letter e—turns out to be a very important discovery. The gradient or steepness of the line with equation $y = e^x$ always has a gradient equal to the y value at that point. This allows calculus to be performed on any exponential graph, which is a graph of equations where the unknown is a power, like $y = 2^x$—important in things like population growth and pandemic situations. Much like π, e turns up in all sorts of areas of mathematics. It is called e after mathematician Leonhard Euler, who famously used it in Euler's identity:

$$e^{i\pi} + 1 = 0$$

This formula, which links five of the fundamental numbers with several key arithmetical concepts, is considered by many mathematicians to be the most profound, elegant and beautiful statement in the field.

You can see a pattern emerging here—it's what mathematicians call a series. The starting value—in this case 100—gets multiplied by the same number (in this case 1.00417) repeatedly to build up the terms in the series. Fortunately, mathematicians have been studying series for a long time, as they are useful for working out the value of things like π, e and various other important values. There is a formula for the total of the series, which I'll couch in terms of your savings: if you invest \$100 each month, at a monthly interest rate of r, for n months, the total value will be,

$$T = \frac{a(r^n - 1)}{r - 1}$$

So, investing \$100 per month for five years (60 months) would give

$$T = \frac{100 \times (1.00417^{60} - 1)}{1.00417 - 1}$$

$$T = \$6,801.30$$

To work out how much you'd need to save each month to make that all-important \$17,899.91, you need to solve this equation:

$$17,899.91 = \frac{a(1.00417^{60} - 1)}{1.00417 - 1}$$

To make life easier, you work out some of the figures on the right-hand side:

$$17,899.91 = \frac{a \times 0.28361431}{0.00417}$$

Now, you multiply both sides by 0.00417:

$$74.6426247 = 0.28361431a$$

Finally, you divide both sides by 0.2836143 to find a:

$$a = \$263.18$$

This value is rounded down to the nearest penny, and will bring you to within 25 cents of the total you require. This is a fair chunk of your current income, but you conclude that it's well worth it for the promise of a million-dollar reward. You laugh at the idea of the look Uncle Ebenezer would have had on his face when he realized you'd met his challenge, but then you think maybe he has done you a favor, and actually got you to think about your finances for the first time. You celebrate by drawing out your weekly earnings and heading down to the local casino.

CHAPTER 14

Prime Location

The message from the aliens has been decrypted! As a senior computer scientist at the SETI Institute, you have been allowed to know the contents of the message and tasked with organizing a response. It appears that their civilization is advanced, friendly and altruistic, and that they are willing to share their technological advances with other civilizations that have made sufficient scientific progress. So, to prove humanity's worth, they have set a challenge. If we can send them a prime number with a 100 million digits, they will send details of several important advances that could reduce our carbon emissions to zero and save the environment from the damage of climate change. Can you find a way to discover such a mon-

strous number?

To start, let's just remind ourselves what a prime number is. All positive whole numbers fall into one of three categories:

- numbers with one factor
- numbers with two factors, itself and 1
- numbers with more than two factors

A factor is a number that divides into the number with no remainder. The number 1 is a factor of every positive whole number, as every number can be divided by 1. For example, 6 can be divided by 1, 2, 3 and 6 with no remainder, meaning it has four factors, putting it firmly in the third category. This third category is called the composite numbers—you'll see why in a minute. The first category is small, as only 1 has one factor. The second category is known as the prime numbers, which are only divisible by themselves and 1. The first few are 2, 3, 5, 7, 11, 13 and 17. It has been proven that there are infinitely many primes. The primes are very special and turn out to be particularly useful for buying stuff on the internet (more on this later).

There is a tasty math fact called the Fundamental Theorem of Arithmetic. It is as important as it sounds. It first says that every positive whole number larger than one is either prime or can be made up by multiplying primes together. So, the composite numbers are so called because they are composed of primes multiplied together. What is more, the theorem also says that each composite number can be made of primes in only one way. As an example, $6 = 2 \times 3$. Or: $123{,}456 = 2 \times 2 \times 2 \times 3 \times 3 \times 173$. There is no other way to make 6 or 123,456 by multiplying primes. This makes prime numbers akin to the DNA of all other numbers.

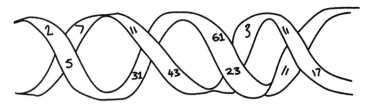

There is no formula or specific way to test whether a number is prime, except by trying to divide the number by prime numbers smaller than it. This makes it difficult to identify large prime numbers and explains why the aliens see it as a test of humanity's development.

Over two thousand years ago, Eratosthenes, an ancient Greek mathematician and head of the fabled library at Alexandria, came up with a method to find primes. Now known as Eratosthenes's Sieve, the process involves making a list of the positive whole numbers. The first prime number is 2, so you mark that as prime and then cross all other multiples of 2 off the list, as they must be composite. Then you move on to the next uncrossed number,

3, and cross off all its multiples that haven't already been crossed out. As you repeat the process, the next uncrossed number must be prime, as you have tested it with all the prime numbers, lower than it. This method is nice, but labor-intensive. If you want to know whether 323 is prime, you have to try dividing it to see whether it is divisible by 2 (no), 3 (no), 5 (no), 7 (no), 11 (no), 13 (no), 17 (yes!). As it is divisible by 17, you know it has at least three factors—1, 17 and 323—and so is a composite number, not a prime number. In fact, 323 = 17 × 19. If the number was composite but had only really large prime factors, it could take a long time to find them.

But does this stuff really matter? If you weren't concerned with getting neat stuff from advanced alien cultures, would anyone care? Well, yes. Prime numbers are used to encrypt traffic on the internet, and the internet is pretty important for a lot of us these days.

Prime numbers are part of the logic behind internet encryption, especially an algorithm called RSA Security, named after its inventors Ron Rivest, Adi Shamir and Leonard Adleman. It is an example of public key cryptography, a system that uses public numbers to encrypt messages and private numbers to decrypt them. The public number is the product (multiplication) of two large prime numbers, and the idea is that it is easy for someone to encrypt a message, but it would take many years of effort to decrypt the message without the private key. The larger the prime numbers you use to make the public key, the more secure your data will be. So, if you're buying things online, prime numbers keep your bank details safe.

Finding a big prime can also be worth a lot of money. The Electronic Frontier Foundation—a nonprofit organization that champions digital privacy—offers a prize of $150,000 for finding the first verified prime number with 100 million digits, so as well as saving the Earth you could earn a bit of money too! But that's a lot of dividing to do, so it would be best if you had a way

The Kardashev scale

Soviet astronomer Nikolai Kardashev was involved in the Soviet Union's own search for alien life. He believed that detectable alien civilizations could—and probably would—be much more scientifically advanced than our own. He came up with a three-stage scale to define them. Type I civilizations would be able to harness the equivalent of all the energy that fell on the surface of their home planet from their sun. Type II would be able to use the entire energy output of a star and Type III would be able to use the energy of an entire galaxy of stars. Humanity is yet to reach Type 1, but using antimatter could help us get there (see chapter 18 for more on this!).

to improve your odds of picking a prime number to check in the first place.

You can eliminate half of the numbers with 100 million digits at a stroke: no even number (apart from 2) can be prime because, by definition, 2 will be a factor. You can get rid of another tenth of the numbers by eliminating any number ending in 5, as these are divisible by 5. There are various other divisibility tricks you can deploy, but even if you eliminate 90 percent of the 100 million digit numbers, that is still an awful lot of very big numbers to check, and even if you had a very fast computer checking those that were left, it could take many years to find and check a prime.

So, before you try a completely brute-force method on the remaining numbers, is there anything that can help? This is where we can employ an idea of Marin Mersenne's, a seventeenth-century French priest. Mersenne was interested in many fields and wrote treatises on music, philosophy and religion, but what we're after stems from his work in mathematics. He was interested in numbers that are one less than a power of two. These numbers are known as Mersenne numbers and have the formula

$$M_n = 2^n - 1$$

To find the first Mersenne number, set $n = 1$:

$$M_1 = 2^1 - 1$$
$$= 2 - 1$$
$$= 1$$

Likewise, for n = 2:

$$M_2 = 2^2 - 1$$

$$= 4 - 1$$

$$= 3$$

You get the idea. This makes the first 11 Mersenne numbers: 1, 3, 7, 15, 31, 63, 127, 255, 511, 1,023, 2,047. Lovely—but how does this help with primes? Well, Mersenne noticed that if n was a prime number, then M_n was also often prime:

Prime	Mp	Prime?
2	$2^2 - 1 = 3$	Yes
3	$2^3 - 1 = 7$	Yes
5	$2^5 - 1 = 31$	Yes
7	$2^7 - 1 = 127$	Yes
11	$2^{11} - 1 = 2,047$	No
13	$2^{13} - 1 = 8,191$	Yes
17	$2^{17} - 1 = 131,071$	Yes

In the above, 2,047 is, in fact, equal to 23 × 89, making it composite. Mersenne primes became the focus of much study, but, as you can imagine, without electronic help these numbers were very hard to check and mistakes were made. Mersenne thought that M_{67} was prime when actually it was composite: $2^{67}-1$ is 147,573,952,589,676,412,927, which is equal to 193,707,721 × 761,838,257,287. This wasn't shown until 1903—over two hundred fifty years after Mersenne's death.

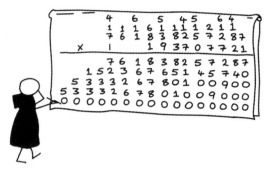

When electronic calculation became available after the Second World War, the relentless task of checking Mersenne primes could be completed much faster and so great strides were made. A team at the University of California verified two new Mersenne primes—M_{521} and M_{607}—in a matter of hours in 1952. To date, fifty-one Mersenne primes have been discovered in total and they fill the top seven slots in the largest primes ever found. Every time a new prime is found, Mersenne or otherwise, it can be used to check for another Mersenne prime, so the system generates new numbers for checking, getting bigger each time. The largest Mersenne prime found to date is $M_{82,589,933}$, which has nearly 25 million digits.

To work out which Mersenne number has at least 100 million digits, you can use the fact that each power of ten has one digit fewer than the power, for example,

$$10^1 = 10 \text{ (2 digits)}$$

$$10^2 = 100 \text{ (3 digits)}$$

$$10^3 = 1,000 \text{ (4 digits)}$$

So, it follows that $10^{99,999,999}$ must have 100 million digits. We need to look at Mersenne numbers to start with that are at least this big:

$$2^n - 1 \geq 10^{99,999,999}$$

It will help the work here to ignore the -1. When you are looking at a number with 100 million digits, taking away 1 makes very little difference. Your equation becomes

$$2^n > 10^{99,999,999}$$

You want to find n, which is a power in this equation, so you need to use logarithms (see Chapter 12):

$$n > \log_2(10^{99,999,999})$$

If you try this on a calculator, you will get an error, as $10^{99,999,999}$ is too large a number for the calculator to cope with. There is a trick we can apply with logarithms, though, due to the way powers work. $10^{99,999,999}$ is 10 multiplied by itself 99,999,999 times. So you could, if you had a lot of time and paper, write it like this:

$$10^{99,999,999} = 10 \times 10 \times 10 \times ... \times 10$$

Why is this important? Well, it allows us to use a rule of logarithms that says, no matter what base you are working in, $\log(a \times a) = \log(a) + \log(a)$. So—

$$\log_2(10^{99,999,999}) = \log_2(10) + \log_2(10) + \log_2(10) + ... + \log_2(10)$$

Altogether, you have 99,999,999 lots of $\log_2(10)$, which is $99,999,999 \times \log_2(10)$. These are numbers the calculator can cope with:

$$n > 99,999,999 \log_2(10)$$

$$n > 332,192,806.2$$

This means you can start your search by finding an existing prime greater than 332,192,806, which, with only nine digits, is relatively tiny. A quick look tells you that M_{31}, discovered by the great Swiss mathematician Leonhard Euler in 1772, has ten digits, being 2,147,483,647. Maybe $M_{2,147,483,647}$ could be a winner?

The last seventeen Mersenne primes found have all been discovered using the Great Internet Mersenne Prime Search—or GIMPS for short. This is a project where participants download a piece of software that uses any redundant processing power of their computer to check for Mersenne primes. Although there are some clever things we can do to help narrow down the field, it is still a very, very big field. You realize that this is not something that the SETI Institute can do by itself. You need to go public and get more people involved. The more people helping, the faster you can find the prime and get help from the friendly extraterrestrials. You tell your bosses at the institute that it's time the world knew the truth—we are not alone.

CHAPTER 15

Handshakes

You are the official photographer for a summit of heads of state, and an expert on catching your subjects at their best. The organizers have asked you to try to capture as many handshake photos as possible. But there's a catch. You have one hour before the last two heads of state arrive. These are a pair of self-important grandstanding buffoons with dubious hairstyles and even more dubious politics, and none of the other one hundred heads of state want to be seen shaking hands with them. When they arrive, you know that all handshaking will stop. Do you have time to take photos of every possible pair of leaders shaking hands before this happens?

Handshake problems are a well-studied area of mathematics, notable for the varied and interesting ways it is possible to solve them. To solve your challenge, it would be useful to know how many photos need to be taken in total. Let's look at small groups of people and see how many handshakes are required.

Anna greets Bob, so one handshake so far:

Carla arrives. She needs to shake hands with Anna and Bob, so the number of handshakes increases by two.

Diego arrives. He needs to shake hands with Anna, Bob and Carla, so the number of handshakes increases by three.

When Edith arrives, she'll need to do four handshakes. And so on. We can see that when, for example, the twenty-fifth person arrives, they'll need to shake hands with the twenty-four people already there. The nth arrival has to shake hands with the $n-1$ people who arrived before them. So, to calculate how many handshakes are necessary for 100 people, you need to work out $1 + 2 + 3 + ... + 98 + 99$.

You could simply spend some quality time with your calculator here, but adding consecutive whole numbers is another old math problem. One way to solve it is with some dotted triangles. If you had five people, you'd need to do $1 + 2 + 3 + 4$ handshakes, which you can show as a triangular array of dots:

You want to be able to get a formula for the number of dots in the triangle. The number of dots in a rectangle is easier to figure out, so, if you add another triangle made up of the same number of dots, you make a rectangle that is 4 dots wide and 5 dots high.

There are $4 \times 5 = 20$ dots in the rectangle, which means there must be $20 \div 2 = 10$ dots in each triangle. Generalizing this, you can see that if the triangle has r rows, the rectangle it makes when doubled up will be r + 1 rows high. This means the rectangle has r × (r + 1) dots in it. To get back to the number of dots in the original triangle, you have to halve this, giving the formula:

$$\text{sum from 1 to r} = \frac{1}{2} r(r + 1)$$

This formula is familiar to mathematicians, as is the anecdote that goes along with it. As a young student in the late 1700s, German math prodigy Carl Gauss was given the task of adding all the numbers from 1 to 100 by his schoolmaster. Legend has it that Gauss spotted this short-cut and solved the problem on the spot instantly, annoying and embarrassing the lazy teacher. Gauss went on to become one of the greatest mathematicians ever.

For n people shaking hands, you need to make r = n − 1, as you have to sum up to one less than the number of people. If you replace r in the formula with n − 1, you get

$$\text{Number of handshakes for n people, } h = \frac{1}{2} (n - 1) (n - 1 + 1)$$

n − 1 + 1 = n

$$h = \frac{1}{2} (n - 1)n$$

So, for 100 heads of state, n = 100:

$$h = \frac{1}{2} \times (100 - 1) \times 100$$

$$h = \frac{1}{2} \times 99 \times 100$$

This equals 4,950 handshakes. To photograph these, budgeting a measly ten seconds per photo, would take 49,500 seconds (s), which is 13 hours (h) 45 minutes (min). So there's no way to take every photo in the hour you have. But the conference center didn't employ you for this kind of can't-do attitude. Let's explore what you can do.

Let's say initially you take one photo of each head of state shaking hands. For 100 people, this would be 50 handshakes, which will take 500 sec to photograph. Then, let's see how many people you could take complete photos of in the time remaining. The time to take the photos is the number of handshakes multiplied by 10 sec, so,

$$\text{Time, } t = 10h$$
$$t = \frac{1}{2}(n-1)n \times 10$$

You simplify this by multiplying the half by the 10 to give 5:

$$\text{Time} = 5(n-1)n$$

World-record-breaking president

On New Year's Day in 1907, President Theodore Roosevelt held an open house at the White House, where members of the public could come and meet their leader. By the time he closed the door, he had shaken hands with 8,513 people. This set a world record for the most handshakes in a day that would not be broken for nearly sixty years. In contrast to our time limit, the record for the longest handshake was set in 2011 by two pairs of handshakers, who agreed to stop after thirty-three hours and three minutes.

$$620 = n^2 - n$$

This type of equation is called a quadratic equation (see page 7), as the unknown—n—is squared. These are not as simple to solve as linear equations, but you can use the quadratic formula if you get the equation into a form where one side of the equation equals zero. To do this for our equation, you subtract 620 from both sides:

$$n^2 - n - 620 = 0$$

Then, you substitute numbers into the quadratic formula, with a = 1, b = -1 and c = -620:

$$n = \frac{-(-1) \pm \sqrt{(-1)^2 - 4 \times 1 - 620}}{2 \times 1}$$

Calculating each part of the equation gives:

$$n = \frac{1 \pm \sqrt{1 + 2{,}480}}{2}$$

This gives a positive solution of n = 25.4 to one decimal place. This means you can take photos of 25 people shaking hands with each other, and it will take $5 \times 24 \times 25 = 3{,}000$ seconds. This leaves you with 100 sec of time in reserve, which you can use to take another ten random photos or mop your sweating brow.

You can represent all these heads of state and their handshakes with a diagram. If you put the 100 heads of state as points on the circumference of a circle and then join the points with lines to represent the handshakes, you get a complex diagram that is known as a mystic rose, perhaps because it looks a bit like the circular rose stained glass windows that you find in old churches. What does seem mysterious is that it appears to make concentric circles and curves, despite the fact that the diagram is made entirely from straight lines. Below is the rose for twenty-five people, generated by Edward L. Platt's Mystic Rose generator at elplatt.com.

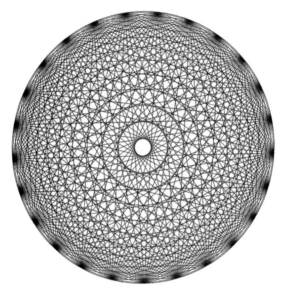

The mystic rose doesn't help with your problem, but it does look very cool.

The organizers are happy with your solution, and you finish off just as a conspicuously large helicopter lands and a cavalcade of big cars pulls up outside the venue.

CHAPTER 16

The Seating Plan

After your escape from Vladivostok, you return home to your fiancée, who happens also to be an international spy. Your big day approaches and the preparations are all in hand. One of the last jobs you need to do is come up with the seating plan for the wedding reception. The problem is, as well as the usual family members that don't get along well together, you and your future wife have a number of friends who are deep undercover and can't be spotted sitting together in case it becomes obvious that they know each other. Is there a mathematical way to come up with a seating plan that guarantees a solution without resorting to trial and error? If you get it wrong, the political implications would be far-reaching and destroy years of careful espionage.

It may seem odd to use mathematics to come up with a seating plan, but seating problems have interested mathematicians for a while now. For example, the Ménage problem was considered in the late 1800s: if you belong to a social circle of married people, how many ways can you sit everyone at a circular table, alternating men and women, and without married couples sitting next to each other?

Despite the simple formulation of the problem, getting an answer is quite tricky. This is partly because at the time it was inconceivable, even for mathematicians working theoretically on paper, for them to do anything but seat all the ladies first. This complicated the mathematics of the situation sufficiently that it took over forty years for the first solution to be found by French mathematician Jacques Touchard. A much simpler non-sexist version of the solution was published in 1986 by Kenneth Bogart and Peter Doyle of the US.

Another seating puzzle known as the Oberwolfach Problem, first posed by German mathematician Gerhard Ringel in 1963, concerns diners at conferences at the Oberwolfach Mathematical

Institute in Germany. It asks how you can seat all the delegates at different-sized circular tables so that each delegate sits with every other delegate only once over the course of a multiday conference. It is known that the number of delegates must be odd, and a few combinations of table sizes have been ruled out, but a general solution to the problem remains to be found. So, if even professional mathematicians through the ages have struggled with this kind of problem, is it possible for you to come up with a solution to your seating problem?

The answer is yes, with a little help from an electronic friend. But first, you need to evaluate all your guests' relationships. You can do this by creating a table of all the guests and giving each of them a score. Give people who don't know each other a score of 0. Give people who do know each other a score of 1. Then, give couples a score of 50 to help ensure they sit together and, vice versa, give people who don't get along with each other a score of -50 to make it clear that they mustn't sit at the same table. You certainly don't want your friends from rival factions of the intelligence community feeling awkward. You could give others scores too, perhaps to represent close or distant family, etc. An example of part of a table of scores might look like this:

	A	B	C	D	E
A	x	50	0	0	1
B	50	x	0	0	1
C	0	0	x	50	-50
D	0	0	50	x	-50
E	1	1	-50	-50	x

A and B are married, and A is the groom's brother. C and D are in a relationship, and D is the bride's cousin. E, the groom's aunt, used to go out with D, and it ended badly. This is called a connection matrix (matrix being a mathematical term for a rectangular array of numbers).

Order matters!

Mathematicians often talk about permutations and combinations. What's the difference? In permutations, the order in which the objects are selected matters, so choosing A, then B, then C counts as being different from C, then B, then A. In combinations, the order does not matter, so A, then B, then C would be the same as C, then B, then A. If I am choosing three books out of five at the shop, the order in which I decide to choose them does not affect my final purchase, so I have a combination. When I get home and am arranging them on my bookshelf, I can position them in different ways—the order I choose would be a permutation. Interestingly, what we call a "combination lock" is very much a misnomer as far as mathematicians are concerned. Order matters—you can't just select the correct numbers; they must be in the correct order too—so really it should be called a "permutation lock."

Once you have scored all your guests against each other, the next job is to look at every possible seating arrangement. Then add up the relative scores of all the people on each table and see which arrangement gives the highest overall score. If all five of the guests in the example matrix opposite were seated together, their score would be 2—the two 50s from the couples A and B and C and D would be canceled out by seating C and D with E. That leaves just the two 1s from E knowing A and B.

To help gauge how much time it will take to compute a viable seating plan, it would help you to know how many seating arrangements are possible. To work this out, you need to know the total number of guests (n) and the number of seats at each table (t). We'll assume that all the tables are the same size and that they all get filled.

You have 104 guests and you're going to seat them at tables of 8. If you filled them at random, there would be 104 choices for the first seat, 103 for the second, 102 for the third, etc. This means that the number of choices for the first table would be

$$104 \times 103 \times 102 \times 101 \times 100 \times 99 \times 98 \times 97 = 10{,}385{,}445{,}095{,}625{,}600$$

That's over 10 trillion permutations, just for the first table! If you carried on through all thirteen tables, you'd have a really huge number—just over 1 followed by 166 zeroes. However, some of these seating plans will be the same, since, if you assign the same 8 people to a table but in a different order, you effectively have the same table. You can arrange 8 people in 40,320 ways, as there are 8 places to seat the first person, 7 for the second and so on:

$$8 \times 7 \times 6 \times 5 \times 4 \times 3 \times 2 \times 1 = 40{,}320$$

Mathematicians have a shorthand for the sort of calculation shown above, called factorial notation. Using this, n! means multiply all the positive whole numbers together, up to and including n. The left-hand side of the calculation above could be written as 8!

So, the first table would have 10,385,445,095,625,600 ÷ 40,320 = 257,575,523,205. A mere 257-odd billion ways! Again, mathematicians have a shorthand for working out this kind of combination problem: $^{n}C_{r}$, where n is the number of things you are choosing from (in our case, wedding guests) and r is the number you are choosing. The formula for $^{n}C_{r}$ is

$$^{n}C_{r} = \frac{n!}{(n-r)! \times r!}$$

Let's test-drive this formula with what we just worked out. We are combining 8 people out of 104, so we set n = 104 and r = 8:

$$^{104}C_{8} = \frac{104!}{(104-8)! \times 8!}$$

This gives

$$^{104}C_{8} = \frac{104!}{96! \times 8!}$$

Evaluating this gives

$$^{104}C_{8} = 257,575,523,205$$

This is as expected.

You can see that, even just for one table, the numbers involved are staggering. What are you going to do? This is where your electronic friend—a computer—comes into play. There are various commercial software packages that can deal with exactly this kind of problem—referred to as linear programming. This can be used for all sorts of applications, such as businesses trying to work out how much of each product they should make to maximize profits. There are even wedding planners who use the software for exactly the predicament you now find yourself in.

In 2012, two academics from Princeton University, Meghan Bellows and Luc Peterson, used this method to work out a seating plan for their 107 guests at 10 tables. Their computer took thirty-six hours to compute the scores for every combination. Apparently, the computer's solution needed only minor tweaks to be acceptable—by the mother of the bride, of course.

Now armed with the wonders of linear programming, you set the computer to try every combination and add up the scores from the connection matrix. The combination or combinations with the highest score would be the best available seating plan for your wedding, keeping undercover agents separated. Happy it's been taken off your hands, you turn to the tricky task of writing a wedding speech that won't spark an international incident.

CHAPTER 17

The Egguation

You are the personal chef of a wealthy African business leader, and you enjoy your job immensely. You spend hours designing nutritious, delicious food using all the marvelous ingredients that Africa has to offer. So you are a bit nonplussed with her latest request—egg and soldiers. She has some family visiting from England, so for her birthday breakfast she asks you to boil an ostrich egg so that the yolk is runny. How long should you boil it for? If the egg is underdone or hard-boiled, you will disappoint your boss and find yourself looking for another job.

Heat is, very loosely, molecules vibrating. The more they vibrate, the more energy they have, so the hotter they are. This vibration can be passed from one thing to another in three ways:

conduction, convection and radiation. Conduction is caused by contact: when I touch my car on a summer's day, the vibration of the molecules on the body of my car makes the molecules in my hand vibrate. Convection is when hot liquids or gases move from one place to another: the radiators in your house should rightly be called convectors as they heat up the air around them, which then moves around the room. Thermal radiation is where photons are emitted by a hot thing, like the sun, and these photons travel and hit things, like the Earth, making their molecules vibrate and therefore warming them.

Different substances transfer heat differently. We know metals get hot quickly, which is why we use them for cooking pots and pans. We know that wood doesn't get hot quickly, and so we frequently attach wooden handles to the metal pots so that we can lift them up.

Heat always flows from hot to cold. What we think of as "getting cold" is actually heat leaving our (hot) bodies into the (cold) environment. The greater the difference in temperature, the greater the flow of heat. If the flow of heat is sufficient, we can get molecules to rearrange themselves—when ice melts or water becomes steam being two examples. Given a nice egg and the means to cause appropriate heat transfer, we can change the molecules in the proteins in the egg to become solid, and if we time it correctly, we can halt the process before the entire egg has solidified—a soft-boiled egg. To get a nice soft-boiled ostrich egg, you need to heat the white to about 63°C all the way through. At this temperature, the yolk should be warm but still runny.

There are several key concepts you need to understand to be able to work out how to heat the ostrich egg.

Specific heat capacity is the amount of energy needed to heat one kilogram of a material by one degree. We saw something like this with the concept of calories in Chapter 9. It takes 4,184 Joules (J) to heat a kilogram of water by one Kelvin (K), which makes the specific heat capacity of water 4,184 J/kgK. Water has one of the highest specific heat capacities of any substance. By contrast, the specific heat capacity of steel is about 490 J/kgK, just over a tenth of that of water, meaning the energy to heat water by one degree will heat steel by ten degrees.

Density was briefly mentioned in Chapters 7 and 10—it is a measure of how large a given mass of a substance is. One meter cubed of water has a mass of 1,000 kg, giving it a density of 1,000

Taking your temperature

Most people use the Celsius scale to measure temperature, which is calibrated on water freezing at 0°C and boiling at 100°C. In the US, we use the Fahrenheit scale, which is calibrated on a salt solution freezing at 0°F and (some say) Daniel Fahrenheit's wife's armpit being at 90°F. Scientists and engineers prefer to use an absolute scale, called Kelvin, which is actually a measure of the heat energy a substance contains. A substance at 100 K has twice as much heat energy than if it were at 50 K. Absolute zero—0 K—means no vibration at all and corresponds to -273.15°C—the theoretical lowest possible temperature. Deep space has a temperature of about 3 K. Each degree on the Celsius scale corresponds to one Kelvin.

kg/m³. Steel's density is about eight times as much, whereas balsa wood's density is around 200 kg/m³. Strangely, the density of ice is 917 kg/m³, explaining why ice floats in water, as less dense substances float in denser ones.

The density of egg white and egg yolk is about the same as that of water—1,038 kg/m³ and 1,032 kg/m³ respectively, explaining why eggs sink when you put them in water.

Conductivity is the main way that heat will flow in our ostrich egg. Thermal conductivity is a measure of how well a substance conducts heat, and it is defined as how much heat energy will conduct through a meter thickness of the material every second for each Kelvin of temperature difference. Insulators have low thermal conductivity: air is a good insulator at 0.026 W/mK, which is why we often use trapped layers of air as insulation. The very best is a vacuum, which does not conduct heat at all as there is nothing to do the conducting. This is a problem for spacecraft, as they can only radiate heat into space, not conduct it. Conductors have high thermal conductivity: copper, a favorite for pots and pans, comes in at around 384 W/mK. Even better is

diamond, which can have thermal conductivity beyond 1,000 W/ mK, but is a tad expensive for cookware.

The equation for boiling an egg involves a lot of complicated geometry as well as thermodynamics. The equation is a bit of a beast:

$$\text{Time} = \frac{c}{\pi^2 k} \times \sqrt[3]{\frac{9M^2 \rho}{16\pi^2}} \times \ln\left(0.76 \times \frac{T_{egg} - T_{water}}{T_{yolk} - T_{water}}\right)$$

Although this equation is very complicated, to use it you need only substitute in the values, which is well within your mathematical capability.

First, ln is a special form of the logarithms that we encountered in Chapter 12. The *logarithme naturel*, which ln stands for, is a logarithm with a base of a number called e; e is a little over 2.7 and, like π, is an irrational number that, as a decimal, carries on forever without repeating itself. Both ln and e have their own button on most calculators, so are easy to use in this formula.

The formula was derived by Dr. Charles Williams, a physicist at the University of Exeter. It was originally intended for chicken eggs. An ostrich egg has the volume of about twenty-four chicken

eggs and enough kilocalories to keep an adult going for an entire day. However, the theory should hold true for this much larger egg, although an ostrich egg's shell is up to six times thicker than a chicken's.

As well as packing a hammer and chisel to open it up, you may wish to extend the cooking time a bit on top of what you calculate to account for the thicker shell. Cookery is very much an art as well as a science! You substitute in the numbers as follows:

- c is the specific heat capacity of the egg white: 3,700 J/kgK
- k is the thermal conductivity of the egg white: 0.34 W/mK
- M is the mass of the egg: 1.4 kg
- ρ is the density of the egg white: 1,032 kg/m³
- T_{egg} is the temperature of the egg before cooking: 20°C, which is 293 K
- T_{water} is the temperature of the water we are cooking the egg in: 100°C, which is 373 K
- T_{yolk} is the temperature we want the yolk to reach: 63°C, which is 336 K

This gives

$$\text{Time} = \frac{3,700}{\pi^2\, 0.34} \times \sqrt[3]{\frac{9 \times 14^2 \times 1,032}{6\pi^2}} \times \ln\left(0.76 \times \frac{293 - 373}{336 - 373}\right)$$

If you type the above into your calculator, you should get

$$\text{Time} = 5,366 \times \ln(1.643)$$

Using that handy natural logarithm button gives

$$\text{Time} = 2,664 \text{ secs (to the nearest sec)}$$

Happy with your calculations, you set about preparing everything. But then, your mercurial boss decides she wishes to have her party at the top of Mount Kilimanjaro. Will your carefully laid calculations still work?

Boiling is a complex process in which substances change from being a liquid to a gas. The temperature at which they do this depends on the atmospheric pressure. In a pan of water, once the water molecules have enough energy to break away from the others in the liquid, they have to fight their way into the air. The more air there is, the harder this is for them to do. At the top of Mount Kilimanjaro, 5,895 m above sea level, air pressure is lower

---

Freezing to death

A common engineering interview question goes something like this: Your plane crashes on the Arctic ice. It is -20°C, and none of the survivors are dressed for this level of cold. Then someone remembers that water can't be below 0°C or it would turn to ice. So, should you all jump in the water to stay "warm"? Seems logical; the water is warmer, after all. The answer is no, and it is because of thermal conductivity. The thermal conductivity of water is over twenty times that of air. Even though the air is colder, water will conduct the heat out of you twenty times more efficiently. So while the temperature difference between your body (37°C) and the air is 57°C, this is only one and a half times the difference between your body and the sea. The higher conductivity of the water overcomes this and makes it—by far—the deadlier option. The flip side of this explains why a swimming pool is lovely on a hot day, even when both the pool and the air are at 27°C/81°F.

---

than at sea level, so it is easier for water to boil. The net effect is that water boils at a lower temperature the higher you go. At 5,895 m, water boils at about 93.7°C. How will this change the cooking time? If T_{water} is now 93.7°C, which is 366.7 K, we get

$$Time = 5,366 \times \ln (1.824)$$

$$Time = 3,225 \text{ secs (to the nearest sec)}$$

This is nearly fifty-four min—ten min more than your first calculation and perhaps not one you'd expect from a measly 6.3 K temperature difference. Fortunately, a helicopter will be involved getting you to the top! You pack your al fresco cooking gear carefully and hope that the vibrations of the helicopter don't scramble the egg before you start cooking it.

You boil the egg for a good hour to take into account the thicker shell and get on with toasting, buttering and cutting soldiers. You successfully achieve a lovely runny, yet warm, yolk, a happy boss and continued employment.

CHAPTER 18

Electric Utopia

You have made the scientific breakthrough that will change the world. Up until now, humanity has had to go to great effort to produce energy. The main form of energy we need is electricity, and you have made a series of discoveries that allow you to convert matter directly into electricity. This will essentially end the need for fuel, which will have a huge impact on the cost of producing and delivering almost everything, from food to smart phones. Climate change will be reversed as you make energy cheap and clean. The financial system, so heavily tied to the price of oil, will be turned upside down, as poverty is eased around the world. This new utopia is all due to your new machine. If the human race currently uses power at a rate of

600 terajoules each year, how much matter will you need to replace this?

$E = mc^2$ is a formula that many can quote, but few can explain. It stems from the work of Albert Einstein and others on relativity. In basic terms, it says that mass and energy are the same thing. Although this three-letter formula appears very simple, it has far-reaching consequences for the way our universe works.

In the formula, E stands for energy in Joules, m for mass in kilograms and c for the speed of light in empty space measured in meters per second. Let's focus on the c for a moment. I mentioned in Chapter 2 that the speed of light in a vacuum is the speed limit for our universe. Nothing can go faster than light does in a vacuum. It does travel very, very fast—just under 300,000,000 m/sec—fast enough for us to feel that light travels around instantaneously. The magnitude of the speed of light has an important role in the equation, so let's look at an example of light's speed to give some context to it.

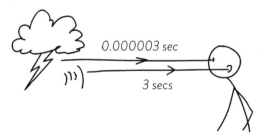

On a stormy night, you see the flare of a lightning bolt in the distance. You start counting "one one thousand, two one thousand, three one thousand," and then the rumble of thunder reaches you. Sound travels in air at about a third of a kilometer per second, so the three seconds you count out mean that the lightning strike happened a kilometer away from you. Light travels in air at about the same speed as it does in a vacuum, so using time = distance ÷ speed,

$$\text{Time} = 1{,}000 \div 300{,}000{,}000$$

$$\text{Time} = 0.000003 \text{ sec}$$

This is three-millionths of a second. There are various arguments for the smallest time interval a human can perceive, but even the quickest of these is in the thousandths of a second. Take something farther away though, and there would be a significant delay in the light reaching you. If Proxima Centauri, the nearest star to the sun, were to explode one day, we wouldn't see it on Earth for four years. When you look at the stars, you are looking back in time. The farthest "star" you are likely to see with the naked eye is, in fact, a galaxy. The Andromeda galaxy is just over 2 million light-years away, so the light you see when you look at

it actually left the galaxy around the time the first *Homo erectus* left Africa.

Einstein reasoned that, if matter and energy are equivalent, then anything that increases an object's energy also increases its mass. There are many types of energy—kinetic energy from movement, gravitational energy from being high up, heat energy from being warm, and electrical energy from being charged up are but a few examples. Most of the time, the increase in mass from these things is negligible. Einstein's formula allows us to calculate the rest mass of an object, which means the energy the object contains excluding all other types of energy it could have.

As an object starts to move, it gains energy, which makes it heavier. Most of the time, for anything moving at less than 10 percent of c—or 30,000,000 m/sec—the increase in mass is negligible. Our fastest spacecraft, NASA's Parker Solar Probe, will reach a top speed of 200,000 m/sec, so it's not something even space explorers need to worry about yet.

The only things that can travel at the speed of light are those that have no mass to start with, like photons. Photons are the smallest unit of electromagnetic radiation and are what make up light. Their rest mass is zero, but obviously they have energy due to their speed.

Einstein's equation means that every kilogram of matter, whether it's water, cheese, a chunk of Pluto or intergalactic space dust, has this much energy:

$$E = 1 \times 300{,}000{,}000^2 = 90{,}000{,}000{,}000{,}000{,}000 \text{ J}$$

This is a lot. Nuclear power stations work on the principle that during fission, where a heavy molecule breaks into two or more lighter ones, the mass of the products is less than the mass of the original particles. The difference in mass is converted to energy, some of which creates heat that we then use for electricity generation. The sun does things the other way around. The fusion process joins two smaller molecules (hydrogen) together to make a larger one (helium). The mass of the helium is not as much as the mass of the two hydrogens combined, so again, energy is released as a consequence, some of which becomes the heat and light that allows all life on Earth.

Even in these nuclear reactions, only a very small percentage of the masses involved get turned into energy. In the nuclear bombing of Hiroshima, only 0.7 grams of uranium was turned to energy, but this is the equivalent of fifteen thousand tonnes of TNT. The sun converts 500 million tonnes of mass into energy every second. That is a lot of Hiroshimas.

Anyway, all this talk of nuclear bombs does not sound very environmentally friendly, and, besides, you want to convert all of the mass into energy. Is this possible?

Well, theoretically, yes. As well as there being matter in the universe, there is also stuff called antimatter. Every matter particle that exists has an antimatter evil twin. Why evil? Well, if a matter particle meets the corresponding antimatter particle, they perfectly annihilate each other—all the mass gets converted to other forms of energy, usually very energetic photons that we call gamma rays. Even if the matter and antimatter are not the same type of particle, annihilation still occurs up to the mass of the lighter particle.

Antimatter bananas

You may have heard that bananas are an excellent source of potassium salts, something your body uses to regulate levels of fluids and hydration. It is a healthy snack and great for preventing cramp for this reason. However, there is a type of potassium called potassium 40 that is very slightly radioactive. When it decays, one of the products is a positron. Antimatter. Only about one in every ten thousand potassium atoms is potassium 40, but this still means that the average banana will produce a positron about once every seventy-five minutes.

Antimatter gets made naturally all the time by reactions that occur from particles decaying. We use this fact in Positron Emission Tomography, a medical imaging technique where a radioactive substance that produces positrons—the antimatter version of an electron—is injected into the body and the resulting energy from the annihilations can be detected by machines that can then form an image of the patient's insides.

Back to solving the world's energy crisis. Six hundred terajoules is 600,000,000,000,000 joules, over 80 percent of which we still get from fossil fuels. The prefix "tera" comes from the ancient Greek work for monster, which seems wholly appropriate in this context. Using Einstein's formula, you can work out how much mass will be required:

$$E = mc^2$$

Dividing both sides by c^2 will make mass the subject of the formula:

$$m = \frac{E}{c^2}$$

Substituting in E = 600,000,000,000,000 and c = 300,000,000,

$$m = \frac{600,000,000,000,000}{300,000,000^2}$$

Squaring the denominator gives

$$m = \frac{600,000,000,000,000}{90,000,000,000,000,000}$$

Simplifying this gives

$$m = \frac{1}{150}$$

So, your invention could power the entire world with a one-hundred-and-fiftieth of a kilogram—less than seven grams—of matter per year. Compare this to the 15 billion tonnes of fossil fuel we get through and it doesn't seem like very much. Such is the power of antimatter!

Why aren't we using it already? There are a couple of reasons. First, it's very hard to store. You can't put antimatter in anything made of normal matter, due to the aforementioned annihilation. Magnetic fields could keep charged particles suspended in a vacuum, but only the very smallest particles are charged. Second, if you weighed all the artificially produced antimatter that humanity has ever managed to make, it would come to a few billionths of a gram.

So, your incredible invention would need to capture and store antimatter, convert all the energy from the annihilations into a useful form of energy, like electricity, and then be able to store it. Quite an invention! Assuming you could make such a thing, it would transform the world and perhaps also allow us to travel to other planets and stars. All because of a little hundred-year-old equation.

Glossary

Acceleration
The change in an object's speed in a given period of time

Arc
Part of the circumference of a circle

Area
The amount of 2-D space a shape takes up

Atom
The smallest possible unit of matter

Ballistics
The study of the movement of projectiles

c
Letter used to denote the speed of light

Calorie
A measure of heat energy usually used in association with food

Celsius
A temperature scale based around the melting and boiling points of water

Circumference
The perimeter of a circle, or the length of the perimeter of the circle

Combinations
Arranging objects in groups where the order of the objects is not important

Composite number
A positive whole number with more than two factors

Concave polygon
A polygon than has one or more interior angles larger than 180°

Conductivity
A measure of a substance's ability to conduct heat

Convex polygon
A polygon whose interior angles are all less than 180°

Denominator
The bottom of a fraction

Density
An object's mass divided by its volume

Diagonal
A straight line that joins two corners of a polygon

Diameter
A line across the center of a circle, or the length of that line

e
A mathematical constant approximately equal to 2.71828

Ellipse
A curved 2-D shape like an egg

Equation
A mathematical statement that links two expressions with an equals sign

Expand
To multiply out a set of brackets

Expression
A set of mathematical symbols, numbers and letters

Factor (n)
A number that divides into another number with no remainder

Factor (v)
To isolate a variable by introducing a set of brackets to an expression

Factorial
The product of all the positive whole numbers less than or equal to the number (e.g., $4! = 4 \times 3 \times 2 \times 1$)

Fahrenheit
A temperature scale based on human body measurements

Floor function
A mathematical process that always rounds down to the nearest whole number

Formula
A scientific theory written mathematically to aid calculations

Friction
The force produced by objects or surfaces rubbing together

g
The acceleration due to gravity at the Earth's surface, equal to 9.81 m/sec^2

G
The gravitational constant that defines how masses attract each other through gravity

Gravitational potential energy
The energy a mass has due to being high up

Gravity
The force that causes masses to attract each other

Hemisphere
Half a sphere

Index / Indices
A number used to denote repeated multiplication

Inequality
Similar to an equation, but compares expressions' relative value

Interior angle
The internal angle at the corner of a polygon

Joule
A unit of energy

Kelvin
An absolute temperature unit that measures the heat energy of an object

Kinetic energy
The energy a mass has due to moving at speed

Kite
A quadrilateral with two pairs of adjacent, equal sides

Light-year
The distance light travels in one year

Logarithm
The inverse of powers and indices

Mass
A measure of how much matter an object is composed of

Matrix
A rectangular array of numbers

Molecule
A particle made from two or more atoms bonded together

Natural logarithm
A logarithm with a base of e

Numerator
The top of a fraction

Orbit
The path a satellite takes

Parabola
The path a projectile takes and the shape of a quadratic graph

Parallelogram
A quadrilateral with two pairs of opposite, equal sides

Parsec
A unit of distance used by astronomers, equal to 31 trillion km

Pentagon
A polygon with five sides

Percentage
A fraction out of 100

Perimeter
The outside of a 2-D shape, or the distance around the outside of a 2-D shape

Permutations
Arranging objects in groups where the order of the objects is important

Polygon
A 2-D shape with straight sides, like triangles and quadrilaterals, etc.

Power
The amount of energy used in a given amount of time, or another word for index

Prime number
A positive whole number with two factors, itself and one

Product
The result of multiplying numbers together

Projectile
An object that has been launched or thrown at a particular speed and angle whose trajectory is only influenced by gravity

Quadratic
An expression or equation where the highest power of the unknown is two

Quadrilateral
A polygon with four sides

Radius
A line from the center of a circle to the circumference, or the length of that line

Resultant force
The net force of all the forces acting on an object

Rhombus
A parallelogram with equal sides

Root
The inverse of raising a number to a power or index

Satellite
An object that orbits another object

Sector
Part of a circle bounded by an arc and two radii

Series
A mathematical sequence where the terms are added together

Simplify
To divide the numerator and denominator of a fraction by a common factor

Specific heat capacity
The amount of energy required to raise the temperature of 1 kg of a material by 1 K

Sphere
A ball

Square root
The inverse of squaring a number

Subatomic particle
A particle smaller than, or that makes up, atoms

Tension
A force caused by pulling

Trajectory
The path of a projectile

Trapezoid
A quadrilateral with one pair of parallel sides

Vacuum
Empty space

Volume
The amount of space a 3-D shape occupies

Weight
The force exerted on a mass by another mass

μ
The Greek letter "mu," used as the coefficient of friction, a measure of the amount of friction between two surfaces

π
The Greek letter "pi," used to denote a circle's circumference divided by its diameter, approximately equal to 3.14

ρ
The Greek letter "rho," used to denote density